P. G. Fake

RADIODETERMINATION SATELLITE SERVICES AND STANDARDS

The Artech House Telecommunication Library

RADIODETERMINATION SATELLITE SERVICES AND STANDARDS

Martin A. Rothblatt

Library of Congress Cataloging-in-Publication Data

Rothblatt, Martin A., 1954–
    Radiodetermination satellite services.

    Bibliography: p.
    Includes index.
    1. Artificial satellites in telecommunication.
2. Artificial satellites in navigation. 3. Mobile
communication systems. I. Title.
TK5104.R68 1987   621.38'0422   87-9141
ISBN 0-89006-239-0

International Standard Book Number: 0-89006-239-0
Library of Congress Catalog Card Number: 87-9141

10 9 8 7 6 5 4 3 2 1

To Dr. Gerard K. O'Neill
for launching RDSS to high orbit

Contents

Preface

In July 1985 the United States Federal Communications Commission (FCC) created a new satellite radio service with tremendous implications for navigation and personal communication technology. The FCC named this set of techniques the *Radiodetermination Satellite Service* (RDSS) and allocated choice frequency bands in the lower microwave region to it. Subsequently, in June 1986, the FCC adopted particular technical and operational standards to govern use of the newly allocated RDSS spectrum. Similar standards were later preliminarily agreed to on a worldwide basis by the International Telecommunication Union (ITU).

This book explains the technical and operational characteristics of RDSS. Chapter 1 provides an overview of the technology, with special emphasis on its unique dependence upon certain portions of the frequency spectrum and on its relation to other mobile communication technologies. Chapter 2 offers a comprehensive discussion of RDSS system architecture. This discussion encompasses design considerations for the overall system as well as its three major components: the space, control, and user segments. Chapter 3 is a working primer on RDSS system management issues. Included here are subsections on the management of traffic in an RDSS system and the control of interference with other RDSS and non-RDSS radio systems operating in the same portion of the frequency spectrum.

Chapter 4 analyzes RDSS applications in a wide variety of industrial and consumer markets. These applications include alternative approaches to the collision-avoidance and emergency-location problems facing aviation, a universal maritime distress and safety system, and an identification of the large market for radiodetermination technology in the land transportation industry. Also identified are the potential for RDSS as a general consumer electronics technology and an interesting array of specialized RDSS applications of interest to governmental authorities.

This book should be of particular value to those involved with advanced communication or electronic navigation technology, whether from policy, engineering, or business perspectives. At present, three RDSS systems are underway in the United States, there is one in Europe, and systems are under study in Australia, China, India, and Japan. The first components of Geostar Corporation's RDSS system in the United States were launched by the European Ariane space rocket in March 1986, and additional RDSS launches are booked, either on Ariane or the NASA space shuttle for nearly every year from 1987 through 1994.

A number of electronics manufacturers are involved in RDSS transceiver production. The first two manufacturers to bring their products to market, Sony Corporation and M/A-COM's Digital Communications Corporation, are gearing up for an international market, while Motorola is developing RDSS transceiver technology for the US government. There are also many opportunities for software vendors, especially in the rapidly growing area of mapping software.

The intent of this book is to provide a number of levels of information so that a wide variety of professionals from spacecraft vendors, electronics manufacturers, software vendors, systems analysts, financial centers, and governmental agencies can quickly access the information that they need to have. With a complex new technology, engineering must talk business and policy must talk operational reality. It is hoped that this book will provide a common frame of reference for all persons involved in RDSS.

Acknowledgments

This book reports on the state of radiodetermination satellite services and standards. The people who were responsible for establishing these services and standards have been largely anonymous for two reasons. First, they are members of corporate or governmental organizations and thus must subsume their identity to that of the organization. Second, most or their work entered the public domain through the regulatory processes of the Federal Communications Commission and the International Telecommunication Union. These individuals and their contributions are given due recognition herein.

Mr. James Laramie and Mr. John Holmes must be credited with putting into writing Dr. Gerard K. O'Neill's conceptualization of RDSS. Messrs. Laramie and Holmes also contributed to the discussion of positioning theory in Chapter 2. Dr. Leslie Snively deserves credit for managing the development of the RDSS control segment and user segment standards presented in Sections 2.3 and 2.4, as well as the RDSS vector geometry and accuracy analysis in Section 2.1.

Dr. William Osborne, Mr. Victor Schendeler, Mr. Dan Lieberman, and Mr. William Cook of Comsat Corporation, Mr. Carl Williams of Railstar Control Technology, and Messrs. Masoud Motamedi, Steve Van Till, and Don Keyser of Geostar Corporation assisted Dr. Snively in his developmental efforts. Much of the credit for the international acceptance of RDSS standards goes to Dr. T. Stephen Cheston, who had considerable assistance from Mr. N.A. Samara and Mr. Alan Rinker of Systematics General Corporation.

Several distinguished individuals foresaw the business potential of RDSS. These people include Mr. David E. Wine, who first conceived of RDSS as a profitable venture; Mr. F. d'Allest, Chairman of CNES, who took the first concrete steps toward the international introduction of RDSS;

Dr. C.J. Waylon, President of GTE Spacenet, whose ideas were instrumental in implementing RDSS in the 1980s. Also critical to the business development of RDSS were Mr. James Wheat's foresight in successfully financing the technology, Mr. Michael Breslin for establishing a true land transportation market, and Mr. Robert Briskman for cost-effective technology implementation via satellite add-on payload design.

The author also wishes to acknowledge the efforts of the FCC's Satellite Radio Branch, especially those of Branch Chief Ron Lepkowski and Senior Attorney Fern Jarmulnek, for providing a public forum for the debate of RDSS standards and for the subsequent establishment of operationally sound parameters and procedures.

The author personally thanks Geostar Corporation for the privilege of serving as an officer since 1985, and to his family for their essential encouragement in and support for this effort. This book never could have been written without the love and understanding of Bina, Eli, Sunee, Gabriel, and Jenesis.

Chapter 1
Description of
Radiodetermination Satellite Service

RDSS is a set of radiocommunication and computational techniques that enables users to determine precisely their geographical position and to relay this and like digital information to any other user. Because these techniques employ geosynchronous satellites, the positioning and messaging capabilities are available over geographical areas of as much as the entire surface area of the earth. The vast majority of RDSS applications concern objects in motion. Hence, in the most general sense, RDSS is a mobile communication technology.

1.1 OVERVIEW Of RDSS

RDSS works by maintaining and manipulating a continuous flow of information between a system control center, geosynchronous satellites, and user terminals, called *transceivers*. The system control center transmits an interrogation signal, at a certain frequency, many times per second to one of three geosynchronous satellites located 35,000 kilometers above the equator. This satellite retransmits that signal at a second frequency to a coverage area on the surface of the earth. The population of user transceivers receives the signal and individual transceivers transmit a response at a third frequency if (a) the user desires a position determination or wishes to send a message, or (b) the signal contains information addressed to that transceiver.

The responses of individual transceivers are received by all three geosynchronous satellites and are retransmitted at a fourth frequency down to the system control center. Due to the varying distances between a user and each of the three satellites as well as the constant speed (velocity of light) at which the signals travel, the control center will receive three identical responses at slightly different times for the signal response of an individual transceiver.

A computer measures round-trip signal transit time by comparing a stored replica of the transceiver's emitted signal with the received signals, then measuring the associated time delay. This time delay, scaled by the velocity of light, is the measurement of the round-trip range from the system control center to the individual transceiver. The round-trip range measurement is converted to three-dimensional coordinates such as latitude, longitude, and altitude.

These position coordinates are then embedded in the continuous interrogation signal sent out to one of the three satellites for retransmission to the earth coverage area. Individual transceivers extract from this interrogation signal those coordinates that are addressed to their ID code. The transceiver then emits a further response to provide the system control center with an acknowledgement that it has received the position coordinates. If the system control center fails to receive the acknowledgement after a period of time, the coordinates are sent again.

Two further points deserve mention at this introductory level of our discussion. First, the same transmission paths that are used to develop positioning information may also be used to send nonpositioning-related alphanumeric information. The amount of such information that may be sent is a variable system parameter. In such cases, the system control center performs a message switching function in addition to a position determination function. For example, a user may want to have either his position or an unrelated message sent to another transceiver. In this case, the user would enter the intended destination transceiver's ID code (or a password that the system control center associates with the ID code) prior to commanding his transceiver to respond to the interrogation signal. Second, it is possible to obtain precision position determination via two geosynchronous satellites, rather than three, if altitude information can be obtained in a different manner. One way that this can be accomplished is by having the transceiver transmit its altitude when it responds to an interrogation signal by incorporating an encoding altimeter into the transceiver. A second way in which this can be accomplished is by accessing a stored digital terrain map at the system control center. Range measurements obtained via two satellites define a line that passes through the earth. A digital terrain map that plots altitude as a function of latitude and longitude can thereby provide altitude information.

In Chapter 2 the positioning methodology described above is rigorously defined as the simultaneous solution of a set of vector equations. The wide range of bias and random errors which can corrupt positioning accuracy are also discussed in Chapter 2 and quantified with respect to the intrinsically differential nature of RDSS techniques. The traffic management issues inherent in a random access acknowledgement-based system such as RDSS are comprehensively addressed in Chapter 3. For now, however, it is appropriate to continue the discussion on a more general level and develop an appreciation for the radio-frequency environment and mobile communication industry within which RDSS exists.

1.1.1 RDSS Territory in the Radio Spectrum

Radio waves are oscillatory changes of electromagnetic fields that propagate at the speed of light. Humanity's efforts to cultivate these fields began in the late 19th century by generating fields that oscillated thousands of times per second (kilohertz or kHz) and today we have reached the point where we routinely generate fields that oscillate many trillions of times per second (terahertz or THz), whereupon the radio waves appear as visible light or lasers. Between these extremes we conquered the very high frequency (VHF) spectrum, consisting of that region from 30 MHz to 300 MHz which is home to television broadcasting, FM radio, and most mobile communication, plus the ultra high and super high frequency spectra, comprising that region from 300 MHz to 30 Ghz which is home to more television broadcasting, cellular telephone systems, microwave ovens, and satellite communications, to name but a few "landlords" in this crowded spectral landscape. It is also helpful to remember that oscillatory frequency varies inversely with wavelength according to the well known formula: wavelength times period of oscillation is equal to the speed of light.

RDSS territory of the radio spectrum exists at two locations in what may be referred to as either the high UHF (thinking in terms of frequency of oscillation) or low microwave (thinking in terms of wavelength) region. The first location is from 16 10–1626 MHz, allocated for transmissions from the user transceivers to the satellites. The second location is from 2484–2500 MIIz, allocated for transmissions from the satellite to the user transceivers. Figure 1.1 graphically shows how the RDSS territory is juxtaposed with respect to other commonly known radio systems in this spectral region.

Refer to Figure 1.1. The International Telecommunication Union divides the frequency spectrum in groups of $3 \times 10^n \mathrm{Hz}$. Due to the relationship: frequency $\times$ wavelength = speed of light = 3×10^8 m, the ITU categorization also divides the frequency spectrum in groups of 10^n m wavelength. For example, the start of the ultra high frequency (UHF) region, 3×10^8, is also the start of the meter-wave region.

Most modern developments in telecommunication technology occur in the UHF and super high frequency (SHF) regions. RDSS allocations occur near the middle of the UHF region. Figure 1.1 expands that portion of the UHF band between 1500 MHz and 2500 MHz, and provides frequency allocation boundaries for the primary users of the various subbands. This subregion is technically desirable for mobile types of services via satellite.

The International Maritime Satellite Organization (Inmarsat) was established to provide satellite communication service to marine vessels. The service is formally known as the Maritime Mobile Satellite Service, and it was the first mobile type of satellite service. The Aeronautical Mobile Satellite Service, known as Aerosat, is a system concept to provide satellite communication service to aircraft, especially when they are beyond the coverage of land-based aeronautical communication systems. Navstar and Glonass are the worldwide satellite navigation systems of the US and Soviet military, respectively. Unlike the switched nature of Inmarsat and Aerosat, these are satellite-to-earth radio broadcasting systems. Hence, there is no user uplink. RDSS, a hybrid mobile communication and satellite navigation service obviously is "right at home" wedged between systems such as Inmarsat, Aerosat, Navstar, and Glonass.

Radio astronomy is a passive radio service in which scientists use ultra sensitive antennas to listen to radio emissions at particular frequency bands from various parts of the universe. The frequency bands required are dictated by physical processes such as atomic energy transition levels, two of which (for the hydroxyl radical) fall within the 1.5–2.5 GHz region. Weather satellites, occupying the 1.670–1.710 GHz region, are well known to the public. Their use of the frequency spectrum is formally known as the Meteorological Satellite Service.

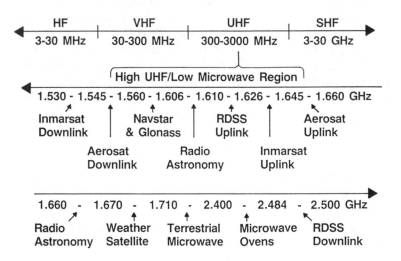

Figure 1.1 Microwave Spectrum

A wide variety of point-to-point terrestrial microwave links, most of which are not more than 100-200 km in length, occupy the 1.710–2.400 GHz region. There are approximately 10,000 of these links in the United States alone. At 2.4 GHz, we enter a region of what is officially called "Industrial, Scientific and Medical" (ISM) uses of the frequency spectrum, wherein consumer microwave ovens are the most common example. Several other bands are also used for such ovens, 915 MHz being the most popular. In a relatively "radio quiet" region just below 2.5 GHz, frequency management officials decided to place the new RDSS downlink.

The reader may recall that four separate radio links are needed in the RDSS architecture. Two of these links connect the satellites to the users and two of these links connect the satellites to the system control center. The two links connecting the satellites to the users are *radiodetermination links* because it is these which are used for determination of position. The two links connecting the satellites to the system control center are considered *feeder links*, or, in spectrum management terminology, "Fixed Satellite Service" links. This terminology means that these links are used to communicate between a satellite and a fixed point on the earth, which in this case is the system control center.

Any fixed satellite service frequency pairs may be used in an RDSS system, and they may be changed from one generation of spacecraft to the next without any effect on the universe of transceivers in the field. There are a large number of frequencies allocated to the fixed satellite service. Table 1.1 lists these frequencies. In the United States, the FCC decided to set aside a particular pair of fixed satellite service frequencies for use by RDSS. These are 6525–6541 MHz for transmissions from the system control center to the satellite, and 5117–5183 MHz for transmissions from the satellite to the system control center.

The reader should note that all frequency bands used by RDSS as outlined above have a bandwidth of 16 MHz with the exception of the 64 MHz bandwidth provided in the 5117–5183 MHz band, which was done to provide enough frequency space for future RDSS satellites to have the downlink capacity to transmit simultaneously several beams of 16 MHz to the system control center. For example, an RDSS spacecraft might cover the United States with four beams, each of which receives transceiver signals from one quadrant of the country. Such a technique generally increases the capacity of the system by as much as a factor of four above that of a single beam covering the entire country. The allocated 64 MHz bandwidth allows the signals received within each of the four beams to be sent to the system control center via a separate 16 MHz channel. If the system design were to employ polarization diversity, then eight 16 MHz channels could be accommodated (four in each of two orthogonally polarized set of channels). Figure 1.2 shows one concept for eight-beam coverage of the United States.

Table 1.1

Worldwide Fixed Satellite Service Frequencies

Frequency (GHz)	Link	Comments
3.40–3.70	Space-to-Earth	C-Band, Mostly International
3.70–4.20	Space-to-Earth	C-Band, Mostly Domestic
4.40–4.80	Space-to-Earth	C-Band, Mostly International
5.11–5.18	Space-to Earth	*RDSS-Related Only in US*
5.85–6.52	Earth-to-Space	C-Band, Mostly Domestic
6.52–6.54	Earth-to-Space	*RDSS-Related Only in US*
6.54–7.07	Earth-to-Space	C-Band, Mostly International
7.25–7.75	Space-to-Earth	X-Band, Mostly Military
7.90–8.40	Earth-to-Space	X-Band, Mostly Military
10.70–12.20	Space-to-Earth	Ku-Band, International and Domestic
12.75–13.25	Earth-to-Space	Ku-Band, Mostly International
14.00–14.50	Earth-to-Space	Ku-Band, Mostly Domestic
18.10–21.20	Space-to-Earth	Ka-Band, Unused
27.50–31.00	Earth-to-Space	Ka-Band, Unused

1.1.2 Zoned Primarily for Positioning Techniques

As regulators prepared to authorize RDSS frequencies, they were faced with a situation of extreme congestion in the needed portions of the spectrum. Technical factors limited the frequencies that RDSS could use to the range of approximately 300 MHz to 3 GHz. These technical factors were a mix of propagation constraints and electronics manufacturing constraints. In this UHF region there were only three bands not already occupied by other systems: the 900 MHz region, the 1610–1626 MHz region, and the pair 1545–1560/ 1645–1660 MHz. The 900 MHz region is hotly contested by land-mobile communication interests such as cellular telephone companies and large private radio operators such as police departments, and thus it was politically untouchable. The 1545–1560/1645–1660 MHz bands were previously zoned for aeronautical voice communication via satellite by international agreement of both the International Civil Aviation Organization (ICAO) and the ITU. Although no system of this kind was in operation, many plans existed to implement such a system, notwithstanding the abortive efforts associated with the Aerosat concept in the early 1970s. A primarily positioning system such as RDSS would clearly violate the zoning limits associated with 1545–1560/1645–1660 MHz.

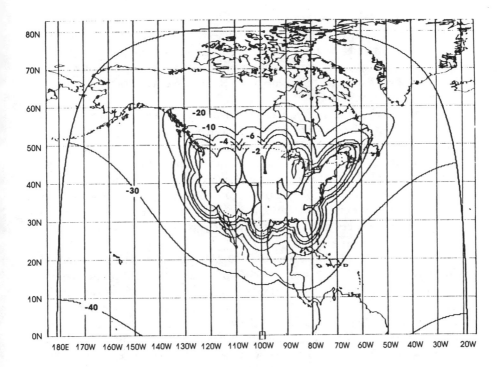

Figure 1.2 USRDSS Central
Note: Gain Contours of Receiving and Transmitting Satellite Antennas at
1618.25 MHz and 2491.75 MHz
Source: US Federal Communications Commission

At 1610-1626 MHz, however, there was an international agreement to
use these frequencies for electronic navigational aids for aviation. The RDSS
system concept clearly fit within this zoning limitation, although the system
concept also enabled electronic navigation to be provided to nonaviation users.
Hence, in July 1985, the FCC determined the RDSS fell "within the purview
of the existing allocation at 1610–1626 MHz."

In fact, the entire 1560–1626 MHz region is generally zoned for satellite
positioning techniques. The sub-band of 1560–1610 MHz is allocated to
Radionavigation Satellite Services. These are services in which numerous
satellites broadcast accurate timing and ephemeris data from which a user
terminal with sophisticated computational capability may determine its geographical position. Examples of such systems are the US Air Force's Global
Positioning System (also known as GPS Navstar) and the Soviet Union's
Global Navigation System (also known as Glonass).

Frequency managers seeking spectral homes for RDSS were still at a loss for an appropriate location for the satellite-to-transceiver link. There were no more empty bands, whether zoned for positioning techniques or not. A difficult decision was made to evict several microwave systems using the 2483–2500 MHz band and to rezone it for RDSS. We will examine why this band was chosen for RDSS.

At several places in the frequency spectrum, permission has been granted for generally unregulated use of what are known as "industrial, scientific and medical" radio devices, or ISM. Examples of such devices are consumer microwave ovens, industrial microwave heating equipment and special medical diagnostic devices, as mentioned above. This type of equipment emits radio waves for noncommunication purposes, and hence generates signals that appear as Gaussian noise. Any communication systems permitted to share a band with these devices must accept any interference that they receive as a result.

One such ISM band exists from 2400 MHz to 2500 MHz, centered on 2450 MHz. Permission had been granted for several hundred point-to-point microwave links and electronic news gathering (ENG) mobile microwave links to share this band on the condition they not complain of ISM interference. An analysis of this band indicated that the fewest number of these links existed in the uppermost portion from 2484 MHz to 2500 MHz. Other parallel analyses indicated that ISM devices would not cause harmful interference to RDSS transceivers unless they were separated by less than about 10 m. The primary reason for this, which will be explored in much greater detail in Chapter 2, is that RDSS transceivers employ *spread-spectrum* modulation techniques. Spread-spectrum is a form of modulation in which the signal itself appears as Gaussian noise due to specialized coding of the information being sent. Because of these factors, regulators were comfortable rezoning 2484–2500 MHz for RDSS, subject to the right of ISM devices to continue to exist.

1.1.3 Zoned Secondarily for Messaging Techniques

While restricting the critical 1610–1626 MHz band primarily for positioning applications, so as to remain in accord with international agreements, regulators did not wish to prohibit use of RDSS for messaging applications. There were three reasons for this. First, messaging was inherently necessary in the RDSS architecture to enable transmission of positioning coordinates from the system control center to a user in the field. Second, there were many obvious cases in which messages other than raw coordinates could be of critical value to safety of life and thereby greatly enhance the value of RDSS. For example, consider the case of an injured cross-country skier trying to summon assistance. RDSS enables not only the transmission of his coordinates to rescuers, but also the ability of rescuers to tell the skier that they are on the way and to advise on self-care until they arrive. Third, and perhaps most importantly, it was clear that the incorporation of a brief alphanumeric message in a response to a satellite interrogation signal did not detract from system capability. In other words, the use of RDSS for limited messaging did not detract from its use as a positioning system. Thus, the regulators ruled that while the primary purpose of RDSS was for positioning, it may be used on a secondary or ancillary basis for messaging. This means that so long as positioning capability is not diminished, the technology may also be used for messaging. A similar situation exists with regard to ancillary use of television and FM radio frequencies for paging and data transmission. The entire bandwidth of a television or FM radio transmission is not needed to broadcast programming. Because of this, television and FM radio licensees are also authorized to provide paging and data transmission services so long as this does not diminish the quality of their broadcast programming.

1.2 RDSS IN THE MOBILE COMMUNICATION ENVIRONMENT

Mobile communication is probably the most diverse and rapidly changing segment of the telecommunication industry. Mobile communication encompasses a wide variety of positioning, data, and voice systems. Some of these are half-simplex, some full-simplex, some half-duplex, and some full-duplex. For positioning, data, or voice applications, there are both land-based and satellite-based systems, and many of these systems may be combined with one another to create hybrid capabilities. There is often more hyperbole than reality, but technologies that were "around the corner" last year are being demonstrated the next year and implemented soon afterward.

The purpose of this section is to focus on where RDSS fits into this rapidly growing industry. Figure 1.3 shows this graphically by plotting coverage area, positioning accuracy, interactivity and vocoded nature, and placing mobile technologies in the portion of the quadrant that they occupy. Table 1.2 is a characteristics matrix, which identifies the unique capabilities of various mobile communication technologies in terms of capacity, capital cost, and user terminal cost.

Refer to Figure 1.3. Cellular telephone, paging, mobile satellite, and RDSS are plotted across a quadrant characterized by parameters of positioning precision, vocoding, coverage area, and directionality. Note that only satellite systems, such as mobile satellite and RDSS, can be truly national. Paging and cellular are "nationwide" only in the sense that they cover specific cities across the nation. Full-simplex RDSS may be operated as either a one-way or a two-way system. Half-simplex paging is limited to one direction of transmission.

Both mobile satellite and RDSS may be vocoded at the voice synthesis level. Only duplex (full or half) mobile satellites may transmit analog voice. Lack of analog voice in RDSS permits five-meter positioning precision, whereas the requirement for analog voice in mobile satellite limits positioning precision. Positioning precision in cellular telephone and paging systems is done by reference to the location of the terrestrial transmitter that relays the telephone call or page.

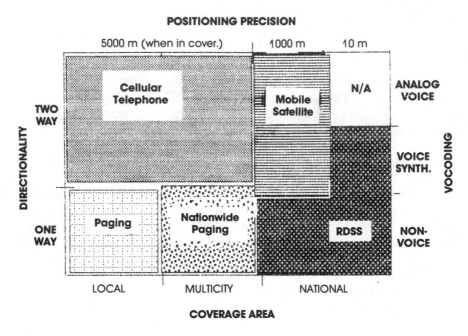

Figure 1.3 Portable and Mobile Communication Systems

Refer to Table 1.2. Average monthly subscription fees assume $10 per call for nationwide mobile satellite service, the 1987 rate for Inmarsat. Monthly fee for "nationwide" cellular assumes $1 for local access and $2 for long-distance charges. Paging charges are based on the 1987 price of 50¢ per "nationwide" page. RDSS charge is Geostar's 1987 flat rate for 24 transmissions per day. Capital cost data are from FCC filings for initial systems. Multibeam satellite systems, not deployable until the late 1990s, would be about three times as expensive.

Table 1.2
Estimated Mobile Communication Technology Characteristics Matrix

	(a) AVERAGE TERMINAL COST		
System Description (Name)	*Capital Cost for 50 Cities*	*Aggregate Capacity with Early 1990s System*	*Average Terminal Cost by 1990*
"Nationwide" Cellular	$1,250,000,000	2,000,000	$1,000
"Nationwide" Paging	$25,000,000	2,000,000	$200
Mobile Satellite	$120,000,000	50,000	$5,000
RDSS	$80,000,000	10,000,000	$500

	(b) MONTHLY SUBSCRIBER FEE		
System Description (Name)	*Capital Cost per Subscriber*	*Maximum Retail Terminal Sales Volume*	*Access Three Times per Day Monthly Fee*
"Nationwide" Cellular	$625	$2,000,000,000	$300
"Nationwide" Paging	$12	$400,000,000	$45
Mobile Satellite	$2,400	$250,000,000	$900
RDSS	$4	$5,000,000,000	$45

1.2.1 Current Array of Mobile Communication Technology

RDSS is but one of many mobile communication technologies available to business and consumers. The key features of RDSS are nationwide coverage, nonvoice transmission, automatic precision positioning capability, and relatively low cost. One RDSS provider, Geostar Corporation, offers its service for two cents per transmission. RDSS hardware vendors Sony and Digital Communications Corporation market their first generation terminals in volume for $2000 to $3000, depending on features. The projection is that service charges will remain fairly constant at two cents per transmission, but that hardware costs will fall to less than $500 in mass production.

1.2.1.1 Private Radio

In terms of number of users, mobile communications technology today is dominated by private half-duplex radio systems. There are approximately 20 million such handsets in use, supported by roughly two million base stations. With such systems, an organization purchases a base station that provides it with two-way radio coverage over a 10 to 50 square-mile area. Typical organizations purchasing such systems are public utilities, public safety departments, and fleet dispatching operators. The base station switches traffic in a broadcast mode among users in the field. Private radio systems are deployed in the 150, 450 and 800 MHz bands.

Studies have shown that the primary use of such systems is to relay locational and status information. RDSS is attractive as a substitute for many of these systems. Value-added enhancements to make RDSS equivalent to private two-way radio include (a) voice synthesis and voice recognition chips, and (b) software-defined private RDSS networks. Voice synthesis and voice recognition techniques will not fully duplicate the analog voice capabilities of private radio, especially for transmissions of more than a few seconds in duration. However, the private radio service is characterized by brief transmissions so as to preserve channel capacity.

In the late 1980s, many efforts are being made to increase the efficiency and utility of private radio systems by using them as a transmission medium for nonvoice information. Motorola, for example, markets an automatic vehicle location (AVL) system in which the position of a mobile unit is determined by using a Loran-C receiver and sent automatically to the company's headquarters via their private mobile radio system. Portable two-way data terminals are also available for sending brief amounts of data over the private radio system. A key distinction between these new systems and RDSS is the nationwide coverage of the latter as opposed to local area coverage for private radio.

1.2.1.2 Cellular Radio

Cellular radio is a public, switched, full-duplex mobile radiotelephone system with worldwide standardized communication protocols, operating in the 800 MHz band. As of 1986, there were approximately 500,000 cellular subscribers worldwide, half of these in the United States. Industry experts expect the US cellular subscriber base to crest at two million *circa* 1990. Because of the standardization of communication protocols and the rapidly growing number of cellular telephone systems, it is possible to extend access to cellular service well beyond the coverage of a "home" or local cellular system. This capability is referred to as "roaming."

Typical purchasers of cellular service are professionals and relatively affluent consumers. Although the cost of the cellular handset has fallen from its introductory price in 1983 of approximately $3000 to 1987 list prices of about $1000 for the least capable units, average cellular service charges have remained constant at about $150 per month. A major drawback to large-scale roaming use of cellular are the high roaming charges imposed by local cellular telephone operators. Operators of large motor fleets are reluctant to install cellular phones because doing so can create an operating cost that is difficult to control as drivers tend to talk longer than necessary when on the road or in congested traffic. Cellular service charges for full-time use in vehicles can exceed $500 per month.

Recently, there have been efforts to incorporate position-reporting devices and two-way data terminals into cellular telephone systems. As with private two-way radio, the intent is to increase the cost-effectiveness of channels designed for voice communication. Unfortunately, the most common positioning receiver, Loran-C, does not work well in cities and the high installation cost of a cellular telephone system generally restricts availability to metropolitan areas. Nevertheless, we can expect the cellular phone network to evolve into a relatively expensive (due to channel scarcity) but ubiquitous transmission medium for mobile and personal communication devices.

RDSS has advantages over cellular for long-distance communication (cellular users must pay normal long-distance telephone charges), positioning accuracy and availability, and the cost of both hardware and service. Cellular's key advantage over RDSS is for voice communication. An interesting synergy between cellular and RDSS is in the area of contacting a roaming cellular user. At present, and for the foreseeable future, when a cellular subscriber (for example, a truck driver) is roaming, his home office has no way to contact him unless he periodically calls in his location. Such calling in becomes very expensive. A truck driver with both an RDSS transceiver and a cellular phone set could have his or her position reported to company headquarters periodically via RDSS at low cost, and perhaps have the cellular phone "locked" to prevent the driver from making calls. Company headquarters could then reach the driver when necessary, and thereby control cellular phone charges. In an emergency, the driver could signal for a call to be made via the RDSS transceiver.

1.2.1.3 Paging

Paging is a half-simplex communication concept in which a mobile user can receive a brief message, but cannot respond in any way via the paging system. The brief message may be a tone or a digital message, such as the display of a phone number or alphanumeric message. Today paging transmissions may be sent over transmitters operating at dedicated paging frequencies in the 150, 450, and 900 MHz bands, or in the sidebands of television and FM radio broadcast transmissions. Paging growth has been especially dramatic in the 1980s, rising from one million subscribers in 1982 to five million subscribers as of 1986. Most industry experts expect 10 million paging subscribers by the early 1990s.

The growth in paging has been primarily due to a reduction in cost for the simplest pagers, from $500 to approximately $100, and in monthly service charges from around $50 to about $15. Another factor enhancing this growth has been the establishment of paging networks, which extend the communication range of a pager from a city to a broad region such as the New York-Washington corridor, and, in the near future via long-haul telephone company trunks, to any of America's largest cities. Such capabilities entail more expensive pagers and paging service charges.

RDSS certainly provides a paging function, but at a significantly higher terminal cost. RDSS transceivers will probably supplant the higher costing pagers, but very cheap consumer-oriented pagers will be a significant feature of the mobile communication marketplace for many years to come. It is interesting to note that the two largest providers of paging service, Mobile Communications Corporation and McCaw Communications, are both authorized by the FCC to establish RDSS systems.

1.2.1.4 Positioning Systems without Communication

There are three non-RDSS positioning systems available for the mobile communication market: Loran-C, Transit, and GPS Navstar. Loran-C is by far the most common, with approximately 250,000 receivers in use. It is operated by the US Coast Guard for general civilian use. Loran-C works at very low frequencies through chains of transmitters that broadcast timing information from which the receivers are able to calculate position with an accuracy of about one mile. Loran-C receivers are available today for as little as $500, most of which are purchased by mariners. More expensive units are found to be most frequently used by general aviation aircraft.

Transit is a US Navy operated satellite positioning system, scheduled for termination in the early 1990s. There are approximately 100,000 Transit receivers in use, each of which determines position based on the Doppler shift of frequencies transmitted by the rapidly moving Transit satellites. A Transit satellite is not always overhead, thus causing delays of up to two hours between position fixes. Recent introduction of large-scale integration (LSI) techniques into Transit receivers has reduced their cost down to approximately $1500. The US government's decision to phase out Transit has truncated general interest in this system.

GPS Navstar is a US Air Force operated satellite positioning system, scheduled for full implementation in 1992. A preoperational test and validation system is already in orbit. With this system, a constellation of 18 satellites in three different 12-hour orbits broadcasts precise timing and ephemeris data to user receivers. The sophisticated computational capabilities in the receivers allow position to be determined with an accuracy of up to 10 m for "secure" users and 100 m for the general public. At present, GPS receivers cost from $30,000 to $150,000, and preoperational service is available only for about four hours per day. However, manufacturers such as Rockwell International project that GPS receivers will decrease in cost to less than $1000 in the 1990s and will find widespread application as part of consumer automobile-navigation systems.

The key distinction between RDSS and the navigation systems described above is that merged communication-navigation capability is available *only* in RDSS. This merged capability has tremendous market significance because (a) most people are on land and (b) people on land generally know where they are. For land users, the desirability of positioning information is to let someone else know their location. Even for maritime and aviation users, however, it is often as important to know their own location as it is to apprise others of their location. Collision-avoidance technology, for example, is dependant on the transmission of one object's position to others nearby. Search and rescue technology also requires the transmission of positioning information.

A second key distinction is that only with RDSS are all position determinations done at a central location, rather than in the user's equipment. A central processing location keeps the RDSS equipment simple, and allows new navigation routines to be made available to all users by simply changing the programming at the system control center. In other words, RDSS lends itself to continual value-added enhancements, whereas the positioning capabilities of Loran, Transit, and Navstar are frozen when a user purchases a receiver.

Finally, the intrinsic positioning accuracy of RDSS is significantly higher than that available from Loran and Transit, commensurate with that which is available only to military users of the Navstar system. This meter-level accuracy is very important to certain classes of applications such as railroad signaling (determining which track a train is on), collision-avoidance systems, and surveying.

1.2.2 Mobile Communication Technology for the 1990s

The 1990s will see the general maturation of the communication technologies listed above, the introduction of GPS Navstar capabilities, a leveling of Loran-C growth, and the termination of Transit. The only new mobile communication technology that can be implemented in the 1990s, due to the time required for regulatory and conceptual development, is one called Mobile Satellite Service (MSS). RDSS will play an important role in determining ultimate market shares for the various mobile communication technologies.

During 1986 the FCC allocated sufficient additional bandwidth to private land-mobile systems and cellular systems to ensure continued growth in capacity until the mid-1990s. Technological improvements in channel utilization, such as narrow-band *amplitude compandered sideband* (ACSB) techniques, provide an additional margin for growth of capacity. In addition, cellular systems are certain to be built in ever smaller communities, leading by the mid-1990s to a situation of nearly ubiquitous cellular coverage. Cellular phone charges are likely to continue averaging around $150 through the 1990s due to demand-driven pricing for high quality voice service. The private two-way mobile communication market will be driven primarily by replacement of existing systems. Replacement systems will probably cost about the same as existing systems. Vendors are bounded at the high end in their system pricing by the availability of competitive systems, such as cellular or RDSS, and bounded at the low end by a captive market of customers who must "use or lose" their FCC licenses to operate a private radio system.

By about 1992 GPS Navstar will supplant Transit as the US government's satellite navigation system. The eventual penetration of GPS into the civilian market will depend on the speed of reduction in price for GPS receivers and the success of RDSS. Government policy is unclear on whether civilian users of GPS must pay a fee. Congress is currently sympathetic to the imposition of such fees and has ordered the Department of Defense to incorporate a fee-charging mechanism into the system architecture. However, at present, there are no GPS user fees for the preoperational system. The final decision on GPS user fees will clearly have an important effect on the relative share of the civilian navigation market enjoyed by GPS.

Loran-C is almost certain to remain on the navigational technology landscape at least until the year 2000. The primary reason for this is the large existing base of Loran-C users and the relatively low (about $50 million per year) cost to the government for maintaining service. Loran-C's accuracy is far inferior to that of GPS and RDSS, but the service is free, and it is difficult to withdraw something that large numbers of people are accustomed to using for free. The current 20% annual growth rate in sales of Loran-C receivers will probably stabilize as target markets in maritime and aviation approach saturation at the same time as RDSS penetrates the land-mobile market.

The only new mobile communication technology foreseen for the 1990s is MSS in the 1545–1560/1645–1660 MHz band pair. MSS is a voice-channelized, full-duplex concept in which transmissions are directly sent between users and a satellite. The capabilities of MSS are similar to those of land-mobile radiotelephone service with the exception of true nationwide coverage for MSS.

A number of regulatory, institutional, and business obstacles will set the pace for introduction of MSS. In the regulatory arena, the United States has required that any use of the MSS band for land-mobile services be subject to the prior rights of aeronautical users so as to satisfy their requirements in that band. This creates an unclear regulatory situation because the aviation community, guided by the FAA, has stated a requirement for exclusive use of the entire band. Technical proposals from prospective MSS operators have demonstrated that use of the band for land-mobile services need not detract from the ability of aeronautical users to satisfy their requirements, but these proposals have not yet been agreed to by the aviation community. Furthermore, the international aviation community, represented by ICAO, is on record as recently as 1985 as being opposed to any land-mobile use of the MSS band. Under present international agreements, the band is restricted to air traffic control types of communication via satellite. In October 1987, a major ITU conference called the World Administrative Radio Conference (WARC) for Mobile Services (MOB-87) will review, and perhaps change, this international agreement.

The major institutional issue involves determining the entity that will be entitled to provide service in the MSS band. The technical characteristics of MSS restrict use of the band to only one system per large geographic area. For example, there could be only one MSS system in North America. In Canada, Telesat Canada Corporation has been selected as an MSS provider, but in the United States some 12 different companies have sought FCC permission to provide the service to land-mobile users, and AirInc, the communications company of the airlines, has requested that it be an exclusive provider of MSS service. The eventual solution probably involves a consortium of some type, but the formation of such a consortium could prove to be a time-consuming process.

The institutional picture becomes a bit more murky when one considers the plans of the Soviet Union, Japan, and Inmarsat to make use of at least a portion of this band to provide service to airlines and maritime vessels over oceanic areas. Such service could cause interference with a domestic MSS system along the coastal United States.

The ultimate issues are in the business arena. The capacity of a current-generation-technology MSS satellite will be limited to approximately 100,000 subscribers due to the need to divide 15 MHz of bandwidth into approximately 1000 15 kHz channels. This is a very small user base for the amortization of approximately $50 million in launching costs, $50 million in satellite construction costs, and at least $25 million in insurance and start-up operating costs. These costs approximately double to launch a redundant system, which is almost certainly necessary to attract manufacturer commitment to mobile satellite phone terminals. For these reasons, US and Canadian MSS providers had been banking on a 1984 NASA commitment to launch the MSS satellites without immediate payment in exchange for a right to use the satellite capacity for experimental purposes. However, the space shutter *Challenger* (January 1986) disaster forced NASA to abandon these plans.

By the mid-1990s, when NASA might again be able to launch an MSS satellite, it will be possible to construct a multibeam MSS spacecraft with the capacity to serve one million or more subscribers. However, by that time, cellular phone service may be so ubiquitous that there will be little market left for MSS.

1.2.3 RDSS Technological Evolutionary Paths

The principal technological evolutionary path for RDSS in the 1990s will be to incorporate *voice synthesis* and *voice recognition* capabilities into user transceivers, and to replace early electronics packages appended to other satellites with large multibeam satellites dedicated to RDSS. Unlike MSS, an operator can begin providing RDSS service by appending inexpensive electronics to another satellite, thereby avoiding the massive capital expenditures associated with a dedicated satellite program. These packages have relatively limited capacity (about one million subscribers), but this is appropriate for the initial development of a new market. By the early 1990s, dedicated RDSS spacecraft can be put in orbit with sufficient capacity to provide RDSS service to many millions of households and businesses. The US government has reserved three space shuttle launching berths for dedicated RDSS satellites in the 1992–1994 time period.

There is likely to be much pressure to incorporate voice capabilities into RDSS. The same RDSS design that enables precision accuracy and high system capacity precludes any sort of full-duplex voice capability. However, voice enhancements to RDSS are readily possible and their implementation is likely.

The most straightforward voice enhancement is voice synthesis, which simply entails incorporating into the transceiver's random-access memory (ROM) the ability to "speak" whatever alphanumeric messages the transceiver receives. Programs are readily available today that perform this function, requiring less than 56 k of memory space. Such programs, for example, allow us to have a personal computer "speak" any amount of text that we type into it.

Voice recognition is a more challenging enhancement for RDSS. This requires the transceivers to "*vocode*" (Voice-Operated Code) information spoken into it, then transmit this digital information in the same manner as it would any information that had been typed into it. The challenge of voice coding lies into human voice variations and language ambiguity. The first step will be to include an EPROM chip in the transceiver, which allows the user to "train" the transceiver to recognize its voice for a limited lexicon. For example, a transceiver user might type in "coming home now" and then speak "coming home now," and press a "*vocode*" key. Thereafter, when he speaks "coming home now" to the transceiver, it will retrieve from memory the digitized message for transmission.

1.3 BRIEF HISTORY OF THE CREATION OF RDSS

A 20-year process is generally involved from the conceptualization of a communication technology to its international implementation and RDSS is no exception. This section completes our introduction by sketching the 20-year path from concept to international service.

1.3.1 ITU Definition of the Service in 1971

The concept of radiodetermination may be traced back to 1912, when seafaring members of the International Telecommunication Union (then called the International Radiotelegraph Union) defined the term as the determination of position, velocity, or other characteristics of an object by means of the propagation properties of radio waves. The definition was a useful reference point for other radio services, but no frequencies were allocated for the explicit purpose of radiodetermination.

In 1971, the member states of the ITU met in Geneva to update the *International Radio Regulations* due to the rapid advances in satellite technology that occured during the 1960s. At the ITU's previous general meeting, in 1959, satellites were still an experimental project and geostationary satellites were not even on the drawing boards. By 1963 satellite communication had become a commercial reality due to the efforts of AT&T and the newly formed COMSAT Corporation. Because of this development, a special meeting of the ITU was held in 1963 (called the Extraordinary Administrative Radio Conference, or EARC) to decide which frequencies should be set aside for satellites. Several frequencies were allocated, but the industry was still too young for a general conceptual framework to fall into place. The purpose of the 1971 meeting, known as the World Administrative Radio Conference for Space Telecommunications (WARC-ST '71), was to establish a general conceptual and regulatory framework for satellite communication.

Several important concepts were established at the 1971 meeting. These concepts can be grouped into three categories. First, the member states agreed that all countries shall have "equitable access" to the geostationary satellite orbit (GSO) and the satellite frequency bands. Second, it was agreed that subsequently launched satellite communication systems must avoid interference with systems launched earlier. Third, a comprehensive set of definitions was adopted for a list of all conceivable satellite services. One such service was the concept of using satellites for determining both the position and velocity as well as other characteristics of an object, and to relay this information elsewhere. The precise wording of the definition was a Chinese contribution, and the concept was named *Radiodetermination Satellite Service* (RDSS). This definition was subsequently incorporated into the ITU's *International Radio Regulations.*

1.3.2 FCC Regulatory Proceedings

Not much was done with the newly defined ITU service between 1971 and 1983. An important ITU frequency allocating conference was held in 1979 (WARC '79), but there were no requests from any country to allocate frequencies for RDSS. At this conference, however, frequencies were allocated to the Aeronautical Mobile Satellite Service (AMSS) and to the Radio Navigation Satellite Service. The AMSS received its allocation at the behest of civil aviation authorities in developed countries. The Radio Navigation Satellite Service received its allocation at the behest of the United States, which intended to use it for the planned Global Positioning System (GPS Navstar).

In late 1982, Dr. Gerard K. O'Neill patented a satellite air traffic control and collision-avoidance system, called *Triad*, due to its reliance on three satellites. Because of FAA reluctance to embrace his concept, O'Neill made efforts to commercialize the concept. These efforts led him to retain frequency management experts to secure the necessary FCC licenses from the US government. The experts grappled with the fact that the invented system did not fit within any existing frequency band allocations, and concluded that a new allocation would be necessary. After some give and take over the system architecture, a fit was made between the Triad concept and the 1971 definition of RDSS. In 1983, O'Neill established Geostar Corporation to obtain the necessary frequency allocation and to commercialize the system.

In the early 1980s, the FCC had already become enamored of marketplace competition and distrustful of monopolies. When faced with the petitions of Geostar, the FCC sought to find a way to authorize the new service but avoid giving a monopoly to one company. After much technical analysis, it decided in 1985 to establish a new Radiodetermination Satellite Service, generally patterned after the Triad system, but required Geostar to amend its system design so as to enable the simultaneous operation of multiple systems. This was accomplished by incorporating the "code division techniques" known as spread-spectrum modulation. In 1986, the FCC formally approved RDSS technical standards based on spread-spectrum modulation and authorized three companies to start building RDSS systems. Provision was also made for other companies to apply for authority to implement RDSS technology. Tables 1.3, 1.4, and 1.5 present the official reallocation of spectrum to RDSS, as it appears in the spectrum management table, with full footnote caveats, for the 1618, 2492, and 5100 MHz bands, respectively.

Table 1.3
US Spectrum Management Table
for 1618 MHz

Government Frequency Allocation (MHz)	Nongovernment Frequency Allocation (MHz)
1610–1626.5 Aeronautical Radio Navigation	1610–1626.5 Aeronautical Radio Navigation

Note: The 1610–1626.5 MHz band is also allocated for use by the Radiodetermination Satellite Service in the earth-to-space direction.

Table 1.4
US Spectrum Management Table
for 2492 MHz

Government	Nongovernment	FCC Use Designators	
Frequency Allocation (MHz)	*Frequency Allocation* (MHz)	*Rule Parts*	*Special-Use Frequencies* (MHz)
2450–2483.5	2450–2483.5 Fixed Mobile Radiolocation	Auxiliary Broadcasting (74) Private Operational- Fixed (94) Private Land-Mobile (90)	2450 ±50 MHz: Industrial, Scientific and Medical Frequency
2483.5–2500	2483.5–2500 Radiodetermination Satellite (Space-to-Earth)	Satellite Communications (25)	—

Note: Stations in the broadcast auxiliary service and private radio services
holding or having applied for a license as of July 25, 1985, may continue
to operate on a primary basis with the Radiodetermination Satellite
Service.

1.3.3 ITU Implementation of the Service in 1987

Although the FCC found a way to authorize RDSS, the commission
recognized that its authorization was not consistent with the international radio
frequency allocations. Efforts were then begun to amend the international
allocations so that the 1610–1626 and 2484–2500 MHz bands were allocated
to RDSS on a worldwide basis. The efforts entail coordination with the ITU,
its International Radio Consultative Committee (CCIR), the International
Maritime Organization (IMO), and the International Civil Aviation Organization ICAO). The IMO and ICAO provide critical information and data to
the ITU on any communication matter of relevance to maritime or aviation.

Table 1.5
US Spectrum Management Table
for 5100 MHz

Government	Nongovernment	FCC Use Designators	
Frequency Allocation (MHz)	Frequency Allocation (MHz)	Rule Parts	Special-Use Frequencies (MHz)
5000–5250 Aeronautical Radio Navigation	5000–5250 Aeronautical Radio Navigation	Aviation (87)	—

Note: The 5117–5183 MHz sub-band is also allocated for space-to-earth transmissions in the Fixed Satellite Service for use in conjunction with the Radiodetermination Satellite Service operating in the 1610–1626.5 MHz and 2483.5–2500 MHz bands. The total power-flux density at the earth's surface shall in no case exceed –159 dBW/m^2 per 4 kHz for all angles of arrival.

The CCIR is the ITU's technical advisory and standards-making body for radiocommunication matters.

Fortunately, in 1983, the ITU had scheduled a future major conference to review all allocations to mobile services with a special focus on those which had application to maritime services. This conference has an official shorthand name of WARC MOB-87, and is scheduled to convene in Geneva in October 1987. In order for RDSS to receive a worldwide allocation at this conference, it would be necessary for (a) the question of RDSS to be placed on the official agenda for the conference, (b) the CCIR to agree on technical standards for RDSS, and (c) a majority of the countries attending the 1987 conference to vote in favor of it.

In early 1985, at the invitation of the ITU's Secretary General, 12 countries asked that frequency allocations for RDSS be placed on the WARC MOB-87 agenda. Subsequently, in July 1985, the ITU's Administrative Council formally voted to place RDSS frequency allocation on the conference agenda. During early 1986, meeting in CCIR working groups, several countries developed agreed-upon technical standards for RDSS. These standards were consistent with those that the FCC had approved for RDSS in the United States.

This effort culminated in the unanimous adoption in 1986 at the XVIth Plenary Assembly of the CCIR of *Report 1050*, the ITU's first set of preliminary standards for RDSS.

A decision will be made at WARC MOB-87 as to whether worldwide frequency allocations should be set for RDSS and, if so, what those frequencies should be. An affirmative answer from this conference will pave the way for RDSS to enter the worldwide mobile communication marketplace. Both the positioning and messaging capabilities of RDSS could then be employed everywhere for the same purposes of safety of life and economic efficiency envisioned for the United States.

In Chapter 2 RDSS system design will be explicated in detail, including its space, control, and user segments. The reader will acquire from this chapter a good understanding of the communication and positioning theory behind RDSS as well as the implementation of such in spacecraft, control centers, and user transceivers.

Chapter 2
RDSS System Architecture

2.1 GENERAL SYSTEM DESIGN

General RDSS system design must be understood from the perspectives of both communications theory and positioning theory. To establish a common framework for these theories, this chapter first reviews the information flows of an RDSS system. Next, we examine the theoretical bases of RDSS communications links — largely an excursion into the realm of spread-spectrum modulation and coding theory. The chapter concludes with an exposition of the geometry and other analytical aspects of position determination in RDSS. An interesting result is the close dependence of RDSS positioning accuracy on the unique features of spread-spectrum modulation. In essence, RDSS general system design results from a convergence of communication and positioning theory.

2.1.1 RDSS Information Flows

The cybernetic maxim "communication is control" applies well to the RDSS system. Information flows are necessarily bidirectional between the user and control segments and, through transitivity, between subelements of the user segment. This means that the user segment can "control," or influence the state of, the control segment and *vice versa*. On the most general level, Figure 2.1 describes the inbound and outbound nature of the RDSS system information flow.

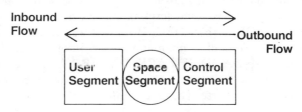

Figure 2.1 RDSS Information Flows

Figure 2.1 shows that information flows along an *inbound* link from the user segment, through the space segment, to the control segment. At the same time, information is flowing along an *outbound* link from the control segment, through the space segment, to the user segment. The RDSS system information flow is a classic example of the Shannon-Weaver dynamic process model of communication. Before looking at the particular contents of these information flows, consider Figure 2.2. This figure reflects user-segment to user-segment information flows with the control segment performing a transitive or *switching* role among inbound and outbound paths.

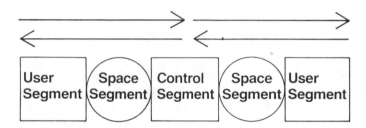

Figure 2.2 RDSS User-to-User Information Flow

A readily understood ordering of the information flow starts with the outbound flow interrogation transmission (let us call it OF-1). This transmission is unaddressed, provides systemic time synchronization, and in essence asks each transceiver whether it requires a position fix or data relay service. Next would come an inbound flow (IF-1) interrogation response, perhaps with a message addressed to another user segment subelement. This, in turn, is acknowledged with a further outbound flow (OF-2), now addressed to one or more particular transceivers. For a simple position determination, this second outbound flow OF-2 is addressed to the transceiver that responded to the unaddressed interrogation OF-1, then acknowledges the transceiver response while providing it with position coordinates. For a message transfer, the one inbound flow OF-1 gives rise to two outbound flows: one acknowledgement to the responding transceiver (OF-2 ACK) and one addressed message to the desired message recipient transceiver (OF-2 ADD). All recipients of this outbound flow acknowledge receipt with a final inbound flow IF-2. We can see that a position determination can entail a minimum of four information flows (OF-1, IF-1, OF-2, IF-2) with full acknowledgements. A message transfer can entail a minimum of six information flows (OF-1, IF-1, OF-2 ACK, OF-2 ADD, IF-2 ACK ACK, IF-2 ADD ACK). Whenever an acknowledgement is not received, the information flow to be acknowledged may be regenerated, of course. The precise number of acknowledgements, the number of unacknowledged information flows repeated before truncation, and the lag time between repeated information flows are all system-dependent parameters.

Table 2.1
Summary of Information Flows

Information Flow	State of Flow and Conceptual Examples
OF-1	Unaddressed interrogation ("Does anyone want to access the system?")
IF-1	Response to interrogation ("I need a position fix" or "I have a message to send")
OF-2 ACK	Acknowledgement of response to interrogation ("Your position is . . ." or "Message is being relayed")
OF-2 ADD	Addressed interrogation ("Message follows . . .")
IF-2 ACK ACK	Acknowledgement ("Received position fix")
IF-2 ADD ACK	Acknowledgement ("Received message")

In the next section of this chapter the parameters of the information flows described above are characterized from the standpoint of communication theory. The purpose of Section 2.1.2 to show that RDSS information flows "work" as a matter of communication engineering. Then, in Section 2.1.3, these same information flows are characterized from the standpoint of positioning theory. We will show that RDSS information flows also result in accurate position determinations.

2.1.2 Communication Theory

The successful transfer of structured packets of energy from one point to another makes electronic communication possible. In satellite communication, the energy is transferred in the form of radio waves. Clearly, the receiving end of the transfer must detect energy sent at the transmitting end in order for the telecommunication *link* to work. This means, for example, that the RDSS control segment must be able to detect energy sent from a portable RDSS user terminal — after a round-trip transit of 72,000 km to and from the geostationary orbit!

There are an infinite number of possible combinations of transmitter and receivers types that could be used to achieve the necessary communication links between the control center and the mobile system users. However, certain desirable characteristics of the system quickly limit the design options. Some of these characteristics are listed below:

- simple, low-cost user terminals;
- high system capacity;
- capability of operating with a minimal restrictions and complexity;
- high precision positioning.

The first objectives contribute to ensuring that RDSS services are available to a large number of users. This large user base also helps to justify the high fixed costs involved in providing the satellite necessary to relay the signals between users and the control facility. The fourth objective is, of course, the primary objective of RDSS.

2.1.2.1 Inbound mobile link

The performance of a communication link can be determined by using a link equation, or *link budget*, to calculate the magnitude of the received signal as a function of the magnitude of the transmitted signals. On the inbound mobile link at 1618 MHz, the signal originates at the user terminal and is transmitted to the RDSS satellites. At this frequency, a typical transmitted power that can be readily achieved in a mobile unit is 40 W or 16 dBW. In evaluating a link equation, it is convenient to express quantities in *decibels* (dB). A power level of 40 W is converted to dB by using the formula $10(\log_{10} 40)$, which equals 16 dBW. The dBW signifies that the value is expressed in decibels relative to one watt. This method of expression is convenient because most of the gains and losses in the link equation are multiplicative factors, which, when converted to dB, can simply be added instead of multiplied. This method also simplifies calculation and understanding of the equation because the various values can easily have magnitudes ranging from 10^7 (70 dBW) to 10^{-23} (–230 dB).

One of the major factors affecting the complexity and cost of the mobile user units is the antenna. An omnidirecitonal, or *isotropic*, antenna is one that transmits equally in all directions. This is the theoretical standard to which we compare all other antennas and to which we measure the *gain* or focusing capability of an antenna. An antenna that focuses all of its transmitted power into a hemisphere, instead of allowing it to be transmitted in all directions as an isotrope would do, effectively doubles the radiated power level in the hemisphere. This factor of two increase, expressed in decibels, equates to 3 dBi, where the *i* signifies that the gain is expressed relative to an isotrope. The combination of the power level of a transmitter (in dBW) and the gain of the transmitting antenna (in dBi) is the *effective isotropically radiated power*, or EIRP. This is a measurement of the transmitting system's total power output.

Directional antennas focus the radiated power more than isotropic antennas, and so they have a higher gain in some directions. The directions over which the antenna has a constant level of high gain is the *beam*, and its width is called the *beamwidth*. High gain antennas, such as parabolic dishes, are generally used in satellite communication. These antennas have relatively high gain, facilitating completion of the communication link. At the same time, high gain antennas have relatively narrow beamwidths, and so they must have some mechanism for pointing the antenna at the intended satellite. This mechanism greatly adds to the cost and complexity of the system, which, coupled with the relatively large size of such an antenna, makes a high gain antenna impractical for hand-held user units or units mounted on small vehicles such as trucks and boats.

Small antennas, having approximately hemispheric beams can be designed to mount flush on a flat surface. These relatively inexpensive antennas have low gain, about 3 dBi maximum, but also have a wide beamwidth of about 140°. This permits the antenna to be mounted on a horizontal surface and "see" any satellite that is more than 20° above the horizon. There is no requirement to steer a hemispheric antenna so that it points at the intended satellite. Such an antenna could have a maximum gain of 3 dBi, with a minimum gain 20° above the horizon of 1 dBi. When combined with the 40 W transmitter described above, this produces a maximum EIRP of 19 dBW and a minimum EIRP of 17 dBW.

As the power radiated from the antenna moves away from the antenna, the power flux density (PFD) of the signal decreases. At any distance, R, from the transmitting antenna, a sphere can be imagined to have a surface area of $(4\pi R^2)$. Because the fixed amount of power radiated by the antenna is spread over larger areas as R increases, the density of that power per square meter must decrease as $1/(4\pi R^2)$, which is commonly called the *spreading* loss. The power flux density of the signal at a distance R from the antenna is EIRP $+ 10\log_{10}(1/(4\pi R^2))$. The spreading loss between a mobile user and a geostationary satellite approximately 42,000 km away is -163.5 dBm^2, bringing the PFD of the minimum transmitted signal to -146.5 dBW/m^2. In addition, there will be a small amount of loss in the link due to the effects of the earth's atmosphere on the transmitted radio waves. At this frequency, this additional loss is no greater than 0.7 dB, reducing the PFD to -147.2 dB.

The antenna on the satellite will receive a certain fraction of the power transmitted by the mobile user, depending on the collecting area of the receiving antenna. A 3.2 m diameter antenna has an area of about 8 m^2. However, because of efficiency factors, the effective collecting area of the antenna is only about $4.4 \, \text{m}^2$ or $6.5 \, \text{dBm}^2$. At the edge of the satellite beam, the effective area would be about 3 dB less, or $3.3 \, \text{dBm}^2$. The PFD of $-147.2 \, \text{dBW}/\text{m}^2$, combined with the effective power-gathering area for a typical 3.2 m diameter antenna of $3.3 \, \text{dBm}^2$, yields a received power level of $-143.9 \, \text{dBW}$ in the satellite receiver.

The information-carrying capability of a communication link is a function of the ratio of the received signal power, or *carrier*, to the noise power density of the receiver. A typical value for the thermal noise in a satellite receiver at 1618 MHz is 600 K (27.8 dBK). The thermal noise power density of a receiver can be calculated as the sum of the noise temperature in dBK and Boltzman's constant ($-228.6 \, \text{dB(W}/\text{Hz-K)}$), which produces a noise power density of $-200.8 \, \text{dBW}/\text{Hz}$ and a carrier-to-noise-density ratio of 56.9 dB.

Report 1050 of CCIR Study Group 8 contains the link budget shown below for a reference RDSS mobile inbound (user-to-satellite) link. This link is based on the same system parameters as discussed in the preceding paragraphs, although the terms in the link budget have been slightly rearranged to conform with general engineering practice. For example, the effective area of the satellite antenna discussed above has been divided into two separate factors. One factor, the effective area of an isotropic antenna ($25.5 \, \text{dBm}^2$), has been included along with the spreading loss in the *free space loss* term. The second part, the gain of the satellite antenna relative to an isotropic antenna (29.0 dBi) is included along with the system noise temperature in the G/T term. G/T is the ratio of the antenna gain to the system noise temperature.

EIRP (1 dBi; 20° elevation)	17.0 dBW
Free-space loss	−189.0 dB
Atmospheric loss	−0.7 dB
Satellite G/T (nominal)	1.0 dBK
C/N_0 (thermal)	56.9 dB-Hz
I_0 (interference from other RDSS)	−2.8 dB
$C/I N_0 + I_0)$ (inbound uplink)	54.1 dB-Hz (2.1)

(Inbound downlink omitted because uplink establishes requirements.)

Information data rate (16 kb/s)	−42.0 dB (b/s)
E_b/N_0 received	12.1 dB
E_b/N_0 required	−10.0 dB

BER of $10^{-5} = 4.6$ dB
SS Implementation loss = 2.0 dB
System margin = 3.4 dB

Margin	2.1 dB (2.2)

where SS = spread spectrum and BER = bit error rate.
Source: CCIR Report 1050 (1986).

As we can see from (2.1) and (2.2) given above, the link budget also includes an allowance for interference from other RDSS users, increasing the noise level by 2.8 dB. The acceptability of any given value of C/N_0 depends on the transmitted signal's data rate. By subtracting the data rate of 42.0 dB (b/s) (or 16 kb/s) from $C/(N_0 + I_0)$, we obtain the ratio of energy per bit to noise density (E_b/N_0). This is the most basic measurement of digital system performance. As we can see from the link budget, the required performance level for these signals is $E_b/N_0 = 10$ dB, leaving a 2.1 dB performance margin.

The values given in the link equation are, of course, subject to some variation. A different sized satellite antenna or an improvement in the gain characteristics of the mobile user's antenna may have a small effect on link performance. A change in the amount of interference due to other RDSS users may also produce a small change in link quality. However, the overall factors that limit link performance are the small antenna size and limited power available to the mobile user. Given these limitations, the maximum achievable data rate is approximately 16 kb/s, as shown above.

2.1.2.2 Inbound Feeder Link

The satellite to control-center portion of the inbound link is an example of standard practice in the fixed satellite service. A baseline example is described below in Table 2.2.

Table 2.2
Baseline Satellite-to-Central Inbound Link for
a 16 MHz Channel in the 6 MHz Band

Satellite EIRP (per user signal)	18.3 dBW
Atmospheric loss	–1.4 1/dB
Path loss	–198.1 dB
Received power P_R (for isotropic antenna)	–181.2 dBW
Earth station G/T	29.4 dB ($1/k$)
k (Boltzmann's constant)	–(–228.6) dB
C/N_0 for inbound downlink to central	76.8 dB-Hz

Source: CCIR Report 1050 (1986).

2.1.2.3 Outbound Feeder Link

The control-center to satellite portion of the outbound communication flow is similarly straightforward, as described in Table 2.3.

Table 2.3
Baseline Central-to-Satellite Outbound Link
for a 16 MHz Channel in the 6 GHz Band

Earth station Transmitted Power	24.0 dBW
Feed loss	–2.5 dB
Earth station gain	51.5 dBi
Earth station EIRP	73.0 dBW
Atmospheric loss	–1.4 dB
Path loss	–200.2 dBW
Received power P_R at satellite (for isotropic antenna)	–128.6 dBW
Satellite G/T	1.5 dB ($1/k$)
k (Boltzmann's constant)	–(–228.6)
C/N_0 for outbound uplink to satellite	98.5 dB-Hz

Source: CCIR Report 1050 (1986).

2.1.2.4 Outbound Mobile Link

Perhaps the most interesting link in an RDSS system is the outbound downlink: the path of transmissions from the satellite to the user. For this link to work well, there must be a sufficiently high ratio of energy per bit to noise density transferred to the RDSS mobile users so as to support a reasonable overall system capacity. As with almost all satellite communication systems, an RDSS system has limited capacity in its satellite downlink because power is most scarce at the satellite.

The basic outbound downlink equation is

Satellite EIRP = Margins – Path losses + k – User G/T + Data rate

With a hemispheric-coverage antenna of 3 dB gain and an assumed effective receiving system temperature of 600 K = –27.8 dB, a user G/T of –24.8 results. Path and atmospheric losses at 2492 MHz total –194.5 dB. As shown in Table 2.4, CCIR Report 1050 sets a baseline link budget with satellite EIRP = 53.6 dBW.

One match of satellite RF power and antenna gain to accomplish this EIRP requirement would be 100 W of RF power and 34 dBi antenna gain. Such a combination could be accomplished on a smaller satellite bus or as a package on a larger (> 1 kW power) spacecraft. For a standard BER of 10^{-5}, the downlink equation implies a maximum data rate of approximately 64 kb/s, which can support an outbound traffic model of 900,000 256-bit messages per hour. It should also be clear that more power with the same antenna gain translates into higher data-rate capacity. Greater antenna gain also enhances capacity, but at the cost of a smaller coverage area on the earth. We shall see later that satellite EIRP has been indirectly fixed by international regulatory policy.

Additional margin is also needed because the link calculations as presented above fail to consider man-made noise, specifically noise caused by other co-channel RDSS systems. Unlike highly directional fixed satellite systems, the RDSS transmissions of users and spacecraft alike normally are within the "look angles" of all other RDSS systems above the horizon. As a rough approximation, received C/N_0 must be reduced by around 1 dB for each RDSS system in operation.

The description in the preceding paragraphs shows the feasibility of establishing links between very small, mobile earth stations and a central control facility by using geostationary satellites.

In order to implement RDSS, however, we must address the previously stated objectives of precise positioning and the ability to serve a large user population. RDSS accomplishes these objectives by the use of *pseudorandom noise codes* (PN codes) to spread the spectrum of the transmitted signals. The PN coding technique reduces system power density, minimizes interference potential, facilitates precision timing, and enables multiple RDSS.

2.1.2.5 Spread Spectrum

The means by which spread spectrum accomplishes coding gain follows naturally from the definition of this type of modulation:

> Spread spectrum is a means of transmission in which the signal occupies a bandwidth in excess of the minimum necessary to send the information; the band spread is accomplished by means of a code which is independent of the data, and a synchronized reception with the code at the receiver is used for despreading and subsequent data recovery. [*IEEE Trans. Communication,* Vol. COM-30, p. 855, May 1982.]

Note here that the fundamental of both spreading and despreading the signal lies in coding the transmission. The coding of a transmission can accomplish through mathematical ingenuity what is virtually impossible to effect by merely increasing the technical performance of transmitting and receiving equipment.

Table 2.4

CCIR Reference Budget for RDSS Outbound Downlink (2491.75 MHz)

Satellite EIRP	53.6 dBW
Free-space path loss	191.8 dB
Atmospheric loss	0.7 dB
User G/T, 1 dBi gain, 20° elevation	-26.8 dB
C/N_0 (thermal)	62.9 dB-Hz
I_0 (provision for 3 other RDSS systems)	-2.9 dB
$C/(N_0 + I_0)$ (total link)	60.0 dB-Hz
Information data rate = 64 kb/s	48.1 dB
E_b/N_0 received	11.9 dB
E_b/N_0 required	9.8 dB
BER of 10^{-5} = 4.6 dB	
Modem loss = 2.0 dB	
System margin = 3.2 dB	
Margin	2.1 dB

Source: CCIR Report 1050 (1986).

Among the different methods of implementing spread-spectrum techniques, the one most relevant to RDSS is *direct-sequence coding*. At the heart of direct-sequence modulation is a form of coding that creates an additional substructure within a stream of digital data. This coding is achieved by *chipping* the outgoing information transmission, modulating the binary stream to give each bit its own internal structure. This forms "bits within bits," which are called *chips*. The internal structure of the information bits — now a series of chips that comprise a bit — is determined by a randomly generated function which is known to both the transmitter and receiver. Because the receiver knows this function, it is possible to decode the sequence of chips and reconstruct the transmitted bit stream. It can therefore be said that spread-spectrum techniques amount to transmitting "extra" information in order to protect the information of real interest.

It is the characteristic of chipping that gives the spread-spectrum modulation technique its name. For electromagnetic waves, the time and frequency profiles of a given signal are intimately related to each other by a mathematical structure known as the Fourier transform. The most significant consequence of this relation is that increasing the temporal fineness of the signal requires that its bandwidth be expanded accordingly. Therefore, because the chipping of the bit stream increases the fineness of its temporal structure, the signal must occupy a wider band than one which is unchipped would require. If coded chipping is regarded as information and frequency bandwidth as a real space, then spread-spectrum has the everyday analog that more information requires more space. Hence, chipping spreads the spectral space.

This analogy does not hold fully because in frequency space different signals can occupy the same space as long as they are sufficiently different to be separable by the user's receiver. What makes spread-spectrum signals "different enough" is the coding embedded in their chip sequences. Within limits determined by permutations on the chip sequence, many such coded signals can coexist in the same band without interference. RDSS systems need this sort of capability in order to accommodate large numbers of simultaneous, nondirectional, and different users.

One of the advantages that the spread-spectrum process provides is that such systems can operate with higher background noise than conventional systems. Bit error rates for all types of transmissions are determined by the signal-to-noise ratio at the receiver. In ordinary transmissions, a bit in error is a lost bit of information, but in spread-spectrum transmissions, a bit in error is only a *chip* lost, not an entire unit of information, because the bit can be reconstructed at the receiver, even if some of the chips are missing. The receiver is capable of doing this because it knows the coded chip sequence and can recognize a bit, even if some of its chips are in error. Because chip or bit errors are most commonly caused by background noise, the receiver's powers of code discrimination allow it to have enhanced performance in noisy applications including operation amid noise generated by other RDSS transmissions.

The amount of processing gain in spread-spectrum modulation is directly proportional to the ratio:

$$\frac{\text{Spread-spectrum bandwidth}}{\text{Minimum bandwidth for bit stream}}$$

The effect of this processing gain is to increase the apparent signal to noise ratio at the point where the bit stream is reinterpreted as data. Because all of the spread-spectrum users within a given bandwidth enjoy the same processing gain, each of them receives the same advantage over competing transmissions as well. For a 64 kb/s outbound data rate and the RDSS 16 MHz allocation, a spreading ratio of 128, or 21 dB exists. This extra 21 dB of margin, in light of the outbound downlink discussion above, ensures that data is faithfully received by the RDSS user.

A final merit of spread-spectrum modulation that is especially important to RDSS applications is its intrinsically high time resolution. Coding or chipping of the digital bit stream increases the fineness of its temporal structure. This characteristic is crucial to RDSS systems because time differences between signals are calculated by comparing the phase differences of two or more incoming signals. By increasing the definition of each incoming signal phase comparisons are made far more accurate.

In a typical spread-spectrum system, there will be at least 100 chips, often many more, occurring within each data bit, each chip taking a very small period of time. A transmitted signal, when finally decoded by the receiving system, will be very precisely matched with the PN code — to a small fraction of a chip. This provides the basis for the precise timing of the received signal required for determining position. A signal with a smaller bandwidth would have a longer chip period and correspondingly produce less accuracy in the timing information. For RDSS systems, a chip rate of about 8 Mcps gives sufficient timing accuracy to provide precise position information. The process of transforming the timing information into usable positioning information will be discussed in the next section.

In summary, the high time resolution required of RDSS signals virtually mandates the use of spread-spectrum coding. Without coding, the actual data rate of the signal would be increased to about 8 Mcps to obtain the necessary time resolution in the transmitted signals. This would also, of course, require an increase in transmitted power level to support the higher data rates. As discussed in the preceding paragraphs, however, data rates of 64 kb/s on the outbound link and 16 kb/s on the inbound link are about the maxima that can be reasonably achieved, given the limits on mobile user unit and satellite technology. The combination of a maximum achievable data rate of 16 to 64 kb/s and the timing precision requirement that the signal be modulated at an 8 MHz rate leads directly to the use of spread-spectrum modulation. The link cannot be achieved without PN coding.

2.1.3 Positioning Theory and Accuracy

The positioning theory behind RDSS will be explicated in the context of determining the position of the aircraft shown in Figures 2.3 to 2.5 below. Coordinates are determined on the basis of the measured arrival times t_1, t_2, and t_3 of the return signals produced in response to an interrogation signal generated by the ground station at time t_0. For convenience, the calculation is carried out in terms of spherical coordinates of the usual form (r, θ, ϕ), where r is the radius measured from the earth's center, θ represents 90° minus latitude, and ϕ represents longitude from the Greenwich prime meridian. Therefore, the respective coordinates of the ground station GS, satellites S_1, S_2, and S_3, and aircraft A may be expressed as follows:

Ground station: $(r_{GS}, \theta_{GS}, \phi_{GS})$
Satellite S_1: $(r_{S1}, \theta_{S1}, \phi_{S1})$
Satellite S_2: $(r_{S2}, \theta_{S2}, \phi_{S2})$
Satellite S_3: $(r_{S3}, \theta_{S3}, \phi_{S3})$
Aircraft: (r_A, θ_A, ϕ_A)

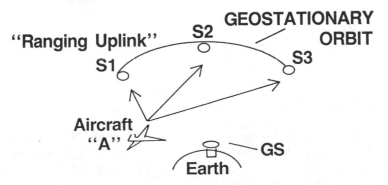

Figure 2.3 Interrogation

All sets of coordinates are known, other than those for the aircraft A. Other necessary quantities that are known or measurable are the interrogation signal transmission time t_0, the response delay T_A of the aircraft, and the response T_S of the satellite repeaters, which is assumed to be the same for all satellites at both RDSS frequencies. We further assumed that all signals uniformly travel at the speed of light c. To the extent that either of these assumptions is inaccurate in a particular case, it is a routine matter to make the appropriate corrections in the equations to follow.

In general, the straight-line distance between two points (r_0, θ_0, ϕ_0) and (r_i, θ_i, ϕ_i) is expressed as follows:

$$d = [(r_i\sin\theta_i\cos\phi_i - r_0\sin\theta_0\cos\phi_0)^2 + (r_i\sin\theta_i\sin\phi_i - r_0\sin\theta_0\sin\phi_0)^2 + (r_i\cos\theta_i - r_0\cos\theta_0)^2]^{1/2} \qquad (2.3)$$

Therefore, the transit time of a signal traversing this distance at the speed of light would be expressed as

$$T = d/c \qquad (2.4)$$

As a convenience, the right-hand side of (2.4) may be expressed in functional notation as follows:

$$T = f[(r_0, \theta_0, \phi_0), (r_i, \theta_i, \phi_i)] \qquad (2.5)$$

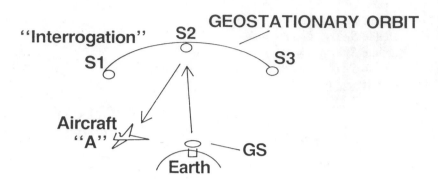

Figure 2.4 Ranging Uplink

We can readily see that the distance measurement, and hence the transit time, will be the same if the coordinates of the two points (r_0, θ_0, ϕ_0) and (r_i, θ_i, ϕ_i) are interchanged in (2.3) and (2.4). In terms of the shorthand functional notation defined above, this means that

$$f[(r_0, \theta_0, \phi_0), (r_i, \theta_i, \phi_i)] = f[(r_i, \theta_i, \phi_i), (r_0, \theta_0, \phi_0)] \qquad (2.6)$$

With reference to Figures 2.3 and 2.4, we can see that the difference between the transmission time t_0 of the interrogation signal from the ground and the time of arrival t_2 at the ground station of the return signal associated with the satellite S_2 will be

$$
\begin{aligned}
t_2 - t_0 = &\, f[(r_{S2}, \theta_{S2}, \phi_{S2}), (r_{GS}, \theta_{GS}, \phi_{GS})] \\
&+ T_S + f[(r_A, \theta_A, \phi_A), (r_{S2}, \theta_{S2}, \phi_{S2})] \\
&+ T_A + f[(r_{S2}, \theta_{S2}, \phi_{S2}), (r_A, \theta_A, \phi_A)] \\
&+ T_S + f[(r_{GS}, \theta_{GS}, \phi_{GS}), (r_{S2}, \theta_{S2}, \phi_{S2})]
\end{aligned}
\qquad (2.7)
$$

If we apply (2.6) and combine terms, this yields

$$
\begin{aligned}
t_2 - t_0 = &\, 2f[(r_{S2}, \theta_{S2}, \phi_{S2}), (r_{GS}, \theta_{GS}, \phi_{GS})] \\
&+ 2T_S + T_A + 2f[(r_{S2}, \theta_{S2}, \phi_{S2}), (r_A, \theta_A, \phi_A)]
\end{aligned}
\qquad (2.8)
$$

Substitution of S_1 and S_3 for S_2 yields two additional equations, which, together with (2.8), form a set of three equations in which the aircraft coordinates (r_A, θ_A, ϕ_A) are the only unknowns. All remaining quantities are either known or directly measurable. These equations may be solved for the coordinates (r_A, θ_A, ϕ_A) by using standard matrix methods. When the solution is completed, the coordinate r_A is converted to the aircraft's altitude above mean sea level by subtracting the radius of the earth r_E, and the coordinate θ_A is converted to the aircraft's latitude by forming the difference ($90°$ minus θ_A). The coordinate ϕ_A is directly equal to the aircraft's longitude.

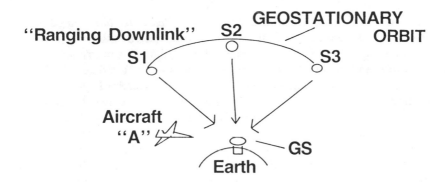

Figure 2.5 Ranging Downlink

2.1.3.1 Accuracy Issues

Positioning accuracy is determined by the magnitude of uncorrectable ranging errors, differential operation considerations, and the geometrical relationship between the satellites and the users. Let us first consider the geometrical relationships.

Geometrical Considerations

Geometrically, the solution to the positioning equations will actually give rise to two possible mirror-image positions for aircraft A, one in the northern hemisphere and one in the southern hemisphere. This ambiguity follows intuitively from the fact that the earth's equatorial plane includes all three satellites S_1, S_2, and S_3, and thus defines a plane of symmetry for the system. This may be visualized by noting that an interrogation signal produced by an equatorial ground station GS at a time t_0 would give rise to the same set of return signal arrival times t_1, t_2, and t_3, for an aircraft located at a given altitude, longitude, and latitude, both north and south of the equator. In reality, however, this ambiguity causes little problem and can be easily resolved in the system's software.

The coplanar geometry of the three satellites introduces a geometric dilution of precision (GDOP) factor of 30–50, depending upon latitude. This means that a one-meter error in ranging results in positioning accuracy of 30–50 m. To address this problem, a two-range approach to positioning is also possible in RDSS. In two-range RDSS, an alternative source of altitude information is substituted for one of the three satellites. Two such sources of altitude information are *encoding altimeters* and *digital terrain data bases*. In the case of an altimeter, altitude information is transmitted to the control center together with the response to the interrogation signal. This information combined with t_0 and the two satellite ranges provides a position determination.

In the case of the terrain data base, consider that a circle of approximately geostationary radii is formed by the intersection of the two spheres that correspond to two satellite range measurements. If we assume that we look at only small portions of that circle, it appears as a straight line. This line must intersect the earth's surface in both the northern and southern hemisphere. The angle of intersection varies with the latitude and longitude of the user. The control center computer solves for the intersection of the bisecting line and a geoid model representing the average height of terrain in the nearby area. A GDOP will still exist for this two-range approach to RDSS, but in this case the GDOP equals the cosine of the angle of intersection, which is about 1.5 to 2.0 in moderate latitudes.

Differential Considerations

RDSS systems may operate in an intrinsically differential mode. This is accomplished by establishing a network of "benchmark" transceivers at widely dispersed, well surveyed locations, then making differential position determination calculations with respect to the known benchmarks. Differential operation eliminates many of the sources of ranging bias errors (ionospheric and tropospheric refraction, system timing errors, delay through all system electronics except the user transceivers) that plague other navigation systems.

Differential corrections can be effected as either a measurement correction or a positioning solution correction. The measurement approach corrects the range measurements before the positioning solution is performed by calculating true ranges to benchmark and subtracting the results from the measured range. This yields the range bias valid for that area, and can be applied to the measured ranges of the various user transceivers in the vicinity of the benchmark before performing the user positioning calculation. Alternatively, we can use the normal positioning algorithm to calculate the position of the benchmark from uncorrected range measurements. In this positioning solution approach, the estimated position is compared to the known position and the difference can be used to correct the positioning solutions of nearby users.

Differential correction will result in a residual error proportional to the distance between the user and the benchmark. This is primarily due to slight nonuniformity of the atmosphere. A widely spaced grid of approximately 300 benchmark transceivers throughout the continental United States (CONUS) would leave residual satellite ephemeris, ionosphere, and troposphere errors on the order of a meter. The most severe atmospheric gradient normally occurs at the "dawn wave," when the sunrise moves across the landscape. Typical values for the dawn wave at the RDSS frequencies are a time of 10 minutes for a variation between nighttime and daytime electron densities. The formula for the dielectric constant k of the ionosphere in the plasma limit at frequency f is

$$k(f) = 1 - (f_p)^2/f^2$$

where f_p is the plasma frequency. The worst case is for the F_2 ionospheric layer, at about 300 km in height, and 1618 MHz frequency. We assume as an extreme that a 50 km slant depth goes in 10 minutes from zero electron density to the maximum midday value, typically 1.5×10^6 electrons per cubic centimeter; the midday value of the plasma frequency is 345 MHz. The resulting value of $k(f)$ differs from unity by 4.5 parts in 100,000. The index of refraction, which is the square root of $k(f)$, therefore differs from unity by 2.3 parts in 100,000. Over a path of 50 km, the delay variation from midnight to midday is therefore 3.8 ns. At a typical latitude of 38°, the dawn wave moves a distance of 210 km in 10 minutes. At a grid spacing of 160 km, corresponding to about 300 benchmarks over CONUS, the ionospheric delay appropriate for each user is calculated by interpolation at an accurately known location between benchmarks, the maximum delay variation of which is 160 km/210 kx $\times$ 3.8 ns = 2.9 ns. If we assume only the grossest correction, e.g., 1/10 of the distance between benchmarks, the ionospheric delay variation can therefore be corrected by the control center to about 0.3 ns, well under one meter of range error.

Range Measurement Errors

If remaining uncorrected, one-sigma errors in range measurements, including residual values after differential correction, would be as listed in Table 2.5.

Table 2.5
Uncorrected Range Measurement Errors

Range Errors	
User transceiver noise	2.8 m
Benchmark transceiver noise	1.4 m
Ephemeris error	0.1 m
Atmospheric delay	0.5 m
Multipath	1.0 m
Root sum square	3.3 m

Table 2.5 (cont'd)

Other Errors
Digital terrain map (nominal)   5.0 m
  (worst case)   30.0 m

User Transceiver and Benchmark Transceiver Noise. Round-trip range measurements are performed by generating a spread-spectrum *pseudonoise* (PN) modulated sequence at the control center, sending it to users via the satellite, and having the user transceiver synchronize to it and track it. This synchronization is used to transmit a second PN code from the RDSS user terminal back to the control center. By comparing the precise phase of the control center receiver's PN code generator to that which generated the transmitted sequence, the control center can then measure round-trip time delay.

To resolve elapsed time to a fraction of a chip, it is necessary to phase-lock to the incoming PN sequence. This sequence can be tracked in various ways, but the *delay-locked discriminator* (DLD) can provide nearly optimal tracking performance. The rms tracking jitter in a noncoherent implementation of the DLD is given as

$$\text{Jitter} = T_c\sqrt{[B_N/(2 \times C/N_0)]}\{1+[2 \times B_{IF}/(C/N_0)]\}$$

where

$$
\begin{aligned}
T_c &= \text{ chip duration;}\\
B_N &= \text{ tracking loop bandwidth;}\\
C/N_0 &= \text{ combined carrier power-to-noise density ratio at the input}\\
&\quad\text{ to the control-center station;}\\
B_{IF} &= \text{ postcorrelator IF bandwidth.}
\end{aligned}
$$

For typical values of C/N_0 = 54 db-Hz and T_c = 80 ns, rms relative errors for various B_N are

B_N (Hz)	*Jitter* (ns)
1000	9.7
300	5.1
100	3.0

B_N values of 1000 can be expected in the user transceiver, and 300 or better can be achieved in the benchmark transceiver. These figures translate into about 2.8 m and 1.4 m of range error, respectively.

Multipath. There are two principal mechanisms for multipath. The first, and most severe, comes from ground reflections when the signal is traveling to the user at very low elevation angles. This form can be virtually eliminated because the spread-spectrum nature of the RDSS design effectively suppresses any signals delayed by more than a chip width. A second form of multipath error is caused by the addition of many slightly delayed signals. This diffuse multipath often occurs due to interactions between the vehicle body and the antenna, and it should not contribute more than one meter of range error.

Altitude Errors. A digital terrain map or altimeter, employed in two-range RDSS, may be thought of as a measurement from a "pseudosatellite" located at the center of the earth. Errors in altitude influence the *absolute* positioning accuracy, but the *relative* accuracy of nearby user is unaffected.

Most terrain is characterized by gradients of less than 3%, which is 158 ft per mile. The density of contour lines on a standard digitized map, with contour lines at 80-ft elevation intervals, therefore averages less than two per mile. For the entire United States, with data points entered every 80 ft along every such contour line, the number of data points required would be about 400 million. We should note that many fewer data points are needed in most areas. Water areas, for example, can be characterized by relatively few numbers because water seeks its own height.

The number of bits required to specify a three-dimensional data point to a fineness of one meter is 56, corresponding to seven eight-bit bytes. The total number of bytes required to characterize the terrain of the United States is less than 2.8 gigabytes. Disk memory units with an access time of 15 ms have been available for several years with that capacity. Interpolation between contour lines can be used to obtain the height at any specific location. Were we to digitize to great precision only mountain ridgelines and shorelines, 500 megabytes would suffice for aviation and maritime safety purposes. Also, we should note that the commercially available US Geological Survey (USGS) digital elevation model contains US terrain elevations spaced three arc seconds apart in a regular array. This translates to 90 m in the north-south direction and about 60 m in the east-west direction at moderate latitudes.

Relative Positioning

It is very important to keep in mind that most users in an RDSS system will be concerned that their reference system be common, not absolute. In short, users will care less where they *actually* are, than they will about where they are *relative* to where they want to be. Landing an airplane, docking a boat, or determining the distance to someplace are all relative problems. RDSS systems can provide highly accurate relative positioning service via *time-difference mapping*. Altitude accuracy issues thus vanish.

With time-difference mapping, round-trip signal times between the control center and points of interest are mapped and recorded through normal system utilization. These signal transit times are corrected by a benchmark measurement to yield a *time difference*. Two such time differences define locations in a unique, repeatable manner. These positions can then be used as the basis for positioning nearby users relative to the time-difference mapped point. Examples of such time-difference mapped points are roadways, railways and airports.

Positioning Accuracy Summary

In summary, positioning accuracy is determined by the magnitude of the ranging error, the error in determining the user's altitude above means sea level, and the geometrical relationship between the satellites and the users. Benchmark transceivers, operating differentially with the RDSS system, can eliminate most errors, including atmospheric variations. Ranging accuracy is a function of the signal-to-noise ratio in the synchronizing circuit, as to which 5 ns is a reasonable figure, corresponding to 1–2 m of range error. This also approximates the fundamental relative accuracy of the system. To achieve an absolute accuracy figure, it is necessary to obtain three satellite ranges, an altimetry input, or a digital terrain map. A GDOP figure of 2 must be employed for the two-range method and a GDOP of 30–50 is needed for the three-range method. This means that in a three-range method, a one-meter ranging error translates into an absolute positioning error of 30–50 m. Alternatively, in the two-range method a one-meter ranging error and a five-meter digital terrain map error becomes a $12 = 2 \times (1 + 5)$ meter error in absolute accuracy of position determination.

2.1.4 Rigorous Vector Analysis of RDSS Positioning Accuracy

Here, we provide a rigorous vector analysis of the accuracy issues previously discussed, as well as several second-order issues. This section may be skipped without loss of continuity but it provides characterization of all necessary parameters for RDSS positioning accuracy.

2.1.4.1 Nominal Vector Equations

From the geometry depicted in Figure 2.6, the vector location of a user R_u can be written in terms of the vectors to satellite A or B ($\mathbf{R}_A$ and $\mathbf{R}_B$) and the range vectors from these satellites to the user, ($\mathbf{P}_A$ and $\mathbf{P}_B$) as

$$\mathbf{R}_u = \mathbf{R}_A + \mathbf{P}_A \tag{2.9}$$

$$\mathbf{R}_u = \mathbf{R}_B + \mathbf{P}_B \tag{2.10}$$

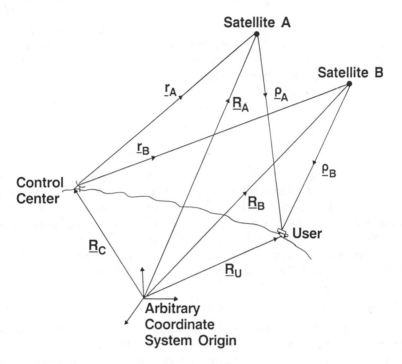

Figure 2.6 Geometry and Vector Definitions

These vector forms are independent of coordinate systems or choice of origin. The vector $\mathbf{R}_u$ is, of course, the unknown position that we seek. Therefore, there are three unknowns to be found and three equations are needed (at least). If we take the dot product of (2.9) with itself and rearrange, we get

$$P_A^2 = R_u^2 + R_A^2 - 2\mathbf{R}_A*\mathbf{R}_u \tag{2.11}$$

where * indicates the vector dot product and a vector symbol that is not in boldface indicates the magnitude of the vector. Similarly, (2.10) gives

$$P_B^2 = R_u^2 + R_B^2 - 2\mathbf{R}_B*\mathbf{R}_u \tag{2.12}$$

Equations (2.11) and (2.12) represent two scalar equations in three unknowns. The *third-dimension* data give the third necessary equation. We can think of this as giving a measure of the magnitude of the user position vector $\mathbf{R}_u$. An altimeter would give a measurement of height above some datum. If the coordinate system were earth-centered and if the earth's radius R_e were known, then the equation would simply be

$$R_u = R_e + h(x,y) \tag{2.13}$$

This altitude $h(x,y)$ is shown as a function of two horizontal position components (or, equivalently, longitude and latitude) to allow for consideration of terrain maps. In that case, a table (or its equivalent by interpolating formulas) would specify the total height as a function of position. In either case, (2.13) represents a third scalar relationship among the three unknown components of $\mathbf{R}_u$. In principle, these nonlinear equations can be solved to yield $\mathbf{R}_u$.

Linearized Error Equations

We need only determine the relations between small errors in the input vector quantities and the resulting small perturbation errors induced in $\mathbf{R}_u$. This is done by replacing every vector in the preceding equations by its nominal value plus a perturbation. The nominal values of all quantities, including user position, are assumed to be known. By neglecting squares of small errors, the relationship between small error contributors and small resulting user-position errors are linear, and thus solved easily. The resulting vector equation for $\Delta\mathbf{R}_u$ can be written as

$$\Delta\mathbf{R}_u = S(n)\Delta\mathbf{p} \tag{2.14}$$

where $\Delta\mathbf{p}$ is the vector of input error parameter values and $S(n)$ is the matrix of sensitivities of $\Delta\mathbf{p}$ upon the components of $\Delta\mathbf{R}_u$. The argument (n) is simply a reminder that these sensitivities depend upon nominal values, and hence will change with user position, satellite position, *et cetera*. If the vector $\Delta\mathbf{R}_u$ is expressed in one coordinate system, then the transformation to local coordinates, such as east, north, and up, is only a matter of multiplying by a transformation matrix $T(n)$, where T depends upon the longitude and latitude of the point in question. The $\Delta\mathbf{R}_u$ vector, as expressed in the new coordinates, can be written as $\Delta\mathbf{R}_u' = T(n) \Delta\mathbf{R}_u = T(n)S(n)\Delta\mathbf{p}$. The new sensitivity matrix in local coordinates is thus $S'(n) = T(n)S(n)$.

Error Sensitivity Results

In Appendix 2.A the sensitivity matrices are tabulated for seven error sources:

- two ranging errors $\Delta\rho_A$ and $\Delta\rho_B$;
- two satellite radius errors ΔR_A and ΔR_B;
- two satellite angular errors $\Delta\theta_A$ and $\Delta\theta_B$;

- (note that strictly speaking two angular errors should be considered for each satellite);
- one third-dimension error Δh.

One set of nominal satellite locations and 12 user locations:

- four with 90°W longitude and with latitudes of 25, 30, 45 and 75°N;
- four with latitude of 40°N and with longitudes of 78, 85, 100 and 125°W;
- four with longitude and latitude fixed at 90°W and 40°N, and altitudes of 0.1, 0.2, 0.3, and 0.4 nmi above the reference datum.

All of the above sensitivity results are presented in earth centered inertial (ECI) and local east, north, and up coordinates. All of the sensitivity values are dimensionless (meters of user error per meter range error, *et cetera*), except for the satellite angular errors, which are in nmi of user error per radian of angular error. This choice of units leads to the large values given in Appendix 2.A.

2.1.4.2 Predicted User Errors and Error Budgets

With sensitivity results available, expected contributions to user position error are computable by

1. specifying values for the error source $\Delta \mathbf{p}$;
2. adopting a philosophy of how the error contributions should combine.

If the errors are all random, independent, and zero mean, then a common practice is to compute the root sum square of the resultant position error; that is,

$$PE = \sqrt{(\text{Error-East})^2 + (\text{Error-North})^2 + (\text{Error-Up})^2}$$

where, for example,

$$(\text{Error-East})^2 = \sum_{j=1}^{7} (S_1, j \ pj)^2 \qquad (2.15)$$

is computed from the first row of the 3 ω 7 sensitivity matrix. The north and up components come from the second and third rows in the same manner.

As a specific example, the results of Appendix 2.A for a user at 90°W and 30°N lead to

$$(PE)^2 = 2.443\Delta\rho_A{}^2 + 1.409\Delta\rho_B{}^2 + 2.418\Delta R_A{}^2 + 1.384\Delta R_B{}^2$$
$$+ 591.3\Delta\theta_A{}^2 \ (PE)^2 \ + 591.5\Delta\theta_B{}^2 + 2.876\Delta_h{}^2 \qquad (2.16)$$

The conversions have been included so that the angle terms are now in feet per microradian (ft/μrad). This means that the angle errors must be expressed

in microradians and all other error values must be in feet. If the reasonable
assumption is made that both ranging errors have the same statistical values,
as do both satellite radius errors and both angle errors, then (2.16) simplifies
to

$$(PE)^2 = 3.852\Delta\rho^2 + 3.802\Delta R^2 + 1182.8\Delta\theta^2 + 2.856\Delta h^2 \qquad (2.17)$$

One approach to developing an error budget is to specify an allowable
position error and assign equal importance to each of the above four contributors. A one-meter system (3.25 ft) would allow each term in (2.17) to have
a value of $3.25^2/4$. This procedure leads to the results of Table 2.6.

Table 2.6
Error Budget

Allowable Position Error	Allowable Contributors			
(meters)	$\Delta\rho$ (ft)	ΔR (ft)	$\Delta\theta$ (μrad)	Δh (ft)
1	0.828	0.831	0.047	0.961
7	5.8	5.9	0.291	6.7
10	8.3	8.3	0.416	9.6
100	83.0	83.0	4.16	96.2

If values for all error contributors are known, then they may be used
to obtain the estimated error in user position. For example, if $\Delta\rho = \Delta R = 10$
ft, $\Delta\theta = 0.4$ μrad, and $\Delta h = 10$ m, then $PE = 63$ ft, or about 19.5 m. Any
combination of error sources can be used to check predicted performance from
either (2.16) or (2.17). If other user locations are to be used, new coefficients
must be obtained by the S matrices in Appendix 2.A.

Another method of using the sensitivity values is to assign numbers to
all error sources except one and then compute the user position error as we
vary the last error source. Results of this type are shown in Figure 2.7 (user
position error *versus* altitude error). Note that in Figure 2.7 three different sets
of errors are used, and three different user latitudes are used in the 10-ft case.
Longitude was also varied, but little change was noted, so we do not plot those
cases.

Another type of result can be obtained by fixing all input error levels
except two and then plotting the locus of those error values that contribute
to constant user position error contours. Figure 2.8 shows the trade-off between

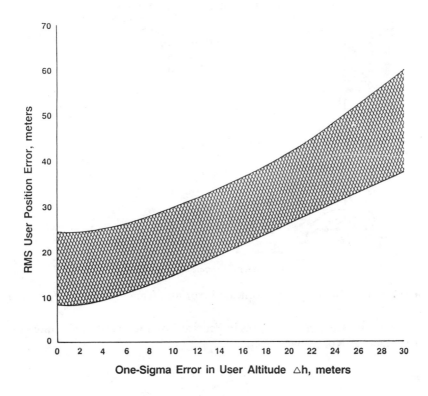

Figure 2.7 Position Accuracy *versus* Altitude and Terrain Map Accuracy (for Nominal Accuracy and User Locations)

third-dimension (i.e., altitude) accuracy and ranging accuracy for the case of very precise satellite position information. The limiting user accuracy, with perfect altitude and ranging, can be seen as 3.9 m. Figure 2.9 shows similar results for a more nominal set of satellite errors. Here, the limiting user accuracy falls to approximately 10 m. Figure 2.10 shows similar kinds of results for the case where errors in satellite radius ΔR are varied along with the errors in ranging, $\Delta \rho$. The limiting value here is 5 m; that is, if the only nonzero erros out of the seven being considered were $\Delta \theta_A$ and $\Delta \theta_B$ of 0.5 μrad each (about 60-ft normal satellite position error), then a 5 m system would result.

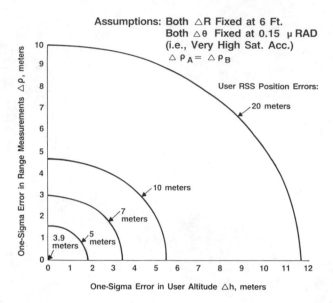

Figure 2.8 Trade-Off between Altitude Error and Ranging Errors (Very High Satellite Accuracy)
Source: Geostar Corporation filings in Federal Communications Commission, General Docket No. 84-690 (1985).

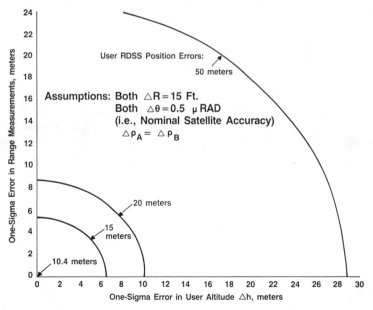

Figure 2.9 Trade-Off between Altitude Error and Ranging Errors (Nominal Satellite Accuracy)
Source: Geostar Corporation filings in Federal Communications Commission, General Docket No. 84-690 (1985).

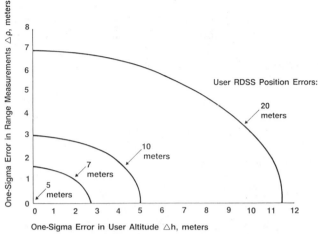

Figure 2.10 Trade-Off between Altitude Error and Ranging Errors (where Satellite Radius Errors Vary with Ranging Errors)
Source: Geostar Corporation filings in Federal Communications Commission, General Docket No. 84-690 (1985).

Bias Errors

Here, we point out that if an error is known to be a bias error, rather than a random error, then there is less justification for using the root sum square (RSS) combination that has been used to this point. Doing so underestimates the effect of the bias by partially burying it along with all random errors. An approach that is often used for known bias errors is to add algebraically the effect to the result obtained from "*RSS*ing" the random errors. The reader should note that the same sensitivity coefficients of Appendix 2.A are still used, but not in the same way. We will give a single example to illustrate. If it is known that there are random components of ranging errors with one-sigma values of 10 ft, and if both satellite positions have random radial errors with one-sigma values of 10 ft and angular errors of 0.4 μ rad, then as found earlier (by *RSS*ing) the user position error would have a one-sigma value of 63 ft. Now, suppose it is known that in addition there is a 3-ft bias error on the range measurement from satellite B due to ionospheric delay. The sensitivity due to this particular range error and geometry is (from Appendix 2.A) –0.77947 ft/ft in the east direction and 1.3552 ft/ft in the north direction. The resulting user position estimate would be biased off to the west by 2.34 ft and to the north by 4.07 ft. An uncertainty superimposed on this offset position of 63 ft still exists. If we desire a single performance value, the vector magnitude $(2.34^2 + 4.07^2)^{1/2} = 4.7$ ft could be added to the 63 ft yielding 67.7 ft, or 20.8 m.

Biases contribute more strongly to final accuracy because they do not tend to average over time. This same behavior, however, makes it easier to calibrate biases out of the system. In the Geostar system, for example, a grid of ground control points is to be used to help calibrate out systematic errors.

2.1.4.3 The Source and Character of Certain Error Contributors

The major error contributors are listed in Table 2.7. Throughout this section we have only discussed certain generic types of errors. Values or ranges of values have been used without any stong justification. In the previous discussion, a distinction was made between bias errors and random errors. In the following subsections we present further discussion of certain errors.

Ionospheric Errors

The ionospheric delay is influenced by a number of factors that change over time: solar and geomagnetic activity, time of day, and season of year. Ionospheric delay is also a strong function of transmission frequency and viewing geometry, and it is highly correlated over sufficiently small spatial regions; that is, it is nearly the same at two nearby spatial points as long as the other parameters are the same. If left uncorrected, delays of up to 50 ns could be expected at midlatitudes of CONUS in the straight-up direction (L band). When slant range (obliquity effects) and GDOP effects are taken into account, errors of up to several hundred feet in user position could result, simply due to this one error contributor. Purely model-based corrections are capable of removing perhaps 70% to 90% of this delay under the best of conditions.

The Navstar GPS program takes advantage of the known dependence of ionospheric errors on frequency by using the so-called *dual-frequency method*. This method uses models and measurements. Measurements of propagation times are made at two sufficiently distinct frequencies, and the results are used with a model to estimate and correct out ionospheric errors. In our example of the Geostar system, only one frequency is to be used, so the dual frequency method is of no utility. However, a large body of experimental research using this method has shown that the vertical delays due to the ionosphere at two points separated by 100 nmi or less have a correlation of 0.95 or more. Thus, the errors would be almost the same at a given time for the same applied obliquity. Another direction of approaching the same conclusion is the published value of 0.065 ns/nmi for the worst known case (*circa* 1958) of ionospheric delay. Thus, if a user and a control point are separated by 160 km (86 nmi), we would expect a difference of about 5.6 ns. If a grid of control points with 160-km spacing were to be used, a user would never be more than half that distance from a control point. Therefore, if the ionospheric delay could be determined at a control point and used as a correction for a nearby user, it appears that the residual error would be less than 2.8 ns. This simple conclusion is not perfect because the obliquity factor would be slightly different at the two points. However, two other factors tend to make the above estimate conservative. First, the separation of the sub-ionospheric points drives the spatial gradient result, not the separation of the points themselves. The rays from a satellite to two earthbound points of separation x will pierce the ionosphere at points separated by less than x. The subionospheric points will be even closer together. Hence, the difference will be less than that predicted by the gradient calculation. Second, the 0.065 ns/nmi value was obtained during 1985, the most active solar year observed to date. Therefore, the 0.065 gradient value is considered to be a maximum.

We have not yet delineated the exact method of using the control points for calibration. It will be considered a little later in Section 2.1.4.4. However, it appears reasonable to assume at this time that proper calibration should allow for removal of most bias types of ionospheric errors, leaving perhaps 2 or 3 ns of residual error. Even if this were doubled when it is *RSS*ed with other ranging errors, its effect will not be dominant. Suppose all other ranging errors amount to 7 ft, then $(7^2 + 6^2)^{1/2} = 9.2$ ft, an increase which does not seem to be too significant in view of Figure 2.8.

Tropospheric Errors

Tropospheric error also contributes a time delay, which is due to refraction of the signal path through a varying density and vapor laden atmosphere at low grazing angles. Tropospheric delay is independent of frequency and if uncorrected can lead to ranging errors of up to 70 ft or more in extreme cases. Tropospheric delay is claimed to be more predictable, however, and thus easier to compensate. This should mean that corrections made at a control point ought to be effective in removing the effect at a nearby user location, as in the case of ionospheric errors.

Multipath Effects

In the Navstar GPS system, multipath effects were of some concern, especially for users operating over water. Multipath is not pursued in detail here, but it is listed only for completeness. Multipath is assumed to contribute a noise on the receiver-measured ranges, and should be included in detailed receiver-error analyses.

Geodesy Errors

When discussing the location of points on the surface of the earth, at least three "surfaces" may be relevant. These are the (1) actual, physical surface, (2) the geoid (the surface that is everywhere normal to the local plumb bob direction and essentially synonymous with mean sea level), and (3) a reference ellipsoid. There are various reference ellipsoids used in practice, which may not have their centers coincide with the earth's center, either because of errors or deliberate attempts at better matching of the geoid over some particular region.

One widely used earth datum is *WGS-72* (World Geodetic System of vintage *circa* 1972). This system defines a reference ellipsoid, then points on the earth's surface are defined in terms of geodetic longitude, geodetic latitude, and height above the reference ellipsoid. This discussion is germane in that it considers positions of satellites and of points on or near the earth. Some of these points are assumed known, such as the location of a control center, a control point, or even a geostationary satellite. A commonly used statement is that any point in CONUS can be located with respect to a central datum by direct survey methods to within 5 vertically and 15 horizontally. (Perhaps better accuracies are available in some classified military situations.) Also, the position of a point *relative* to some other nearby point clearly can be determined more accurately. However, if this relative positional data must be converted to some absolute datum, the overall accuracy is corrupted toward the 5 m and 15 m figures. Therefore, a known point is not really known exactly. If everything used is expressed in the same datum, local to CONUS, then this may not be a consideration. However, quite often satellite positions are expressed in a more absolute coordinate frame, such as ECI. If such is the case, then the uncertanties in the absolute locations apply.

Terrain Mapping Errors

A terrain map can take various forms. All of the mapping forms basically try to describe the physical height of points on earth above some datum, such as the geoid or a reference ellipsoid. It matters little whether the map is a polynominal fit, a plane figure fitted to three corner points, or just a very fine grid of tabulated heights. None will capture every undulation in the earth's surface. However, two considerations arise: how well known are the control points in the datum being used, and how well do the interpolating methods fit the surface in between? Various accuracies are quoted for terrain maps, but it seems that the bound on accuracy is probably the 5 m mentioned above. How much degradation occurs from this is determined by data storage capacity. The only intention here is to justify the 1 to 30 meter range that was used in Section 2.1.4.2 (Figures 2.7 to 2.10).

Satellite Ephemeris Errors

The errors in knowledge of satellite position at geosynchronous altitude can be kept to within about 10 ft in the radial direction and about 50 ft each in cross-track and in-track directions. These errors are one-sigma values that can be obtained from trilateration, which is three-station ranging with reasonably well spaced ground stations. Notice that 50 ft at approximately 19,000 nmi subtends an angular error of about 0.4 μrad. If the two 50-ft errors are combined into a total horizontal error of 70.7 ft, the associated angular error is about 0.6 μrad. These levels of accuracy assume reasonably frequent updating of position. A recent study on the potential for collision of satellites in the geosynchronous band stated that trilateration is able to give position predictions, good for a two-week period, with errors on the order of .054 nmi, i.e., 328 ft. This translates into an approximate angular error of 2.9 μrad. This range of values provides bounds for what we call *satellite ephemeris errors*. As is clear from the error analysis equations of Appendix 2.A, the results here are based on a single angular error for each satellite, rather than two (in-track, cross-track). The single angle used was neither in-track nor cross-track, but rather a combination in the plane formed by the satellite and user position vectors. We assume here that this is the dominant error-producing orientation.

Algorithmic and Numeric Errors

The set of three simultaneous nonlinear equations (2.11), (2.12), and (2.13) was solved numerically by using single precision arithmetic and an interative algorithm. For this simple test, $h(x,y)$ was merely a constant. The data for ρ_A and ρ_B *et cetera* were generated by using exact values of the satellite and user positions and we added no errors of any kind. The process converged to an answer for user positions that differed from the true input position by 2 ft to 8 feet. These errors could be due to poor problem formulation. (Note that the process involves taking small differences of large numbers.) Errors might also be due to lack of precision in the algorithm and its convergence criterion.

2.1.4.4 Use of Control Points to Eliminate Biases

Assume that at least three control points are available and that their locations $\chi 1$, $\chi 2$ and $\chi 3$ are known precisely. Then, modification of (2.9) gives

$$\mathbf{R}_A + \rho_{Ai} = \chi i$$

where i can take on values of 1, 2, or 3. If an unknown bias $\mathbf{b}_A$ exists on $\mathbf{R}_A$

and another bias $\mathbf{b}_i$ exists on ρ_{Ai}, then

$$\mathbf{R}_A + \mathbf{b}_A + \rho_{Ai} + \mathbf{b}_i = \chi_i \tag{2.18}$$

It is clear that distinction between $\mathbf{b}_A$ and $\mathbf{b}_i$ will not be possible because they always appear as a sum, no matter how many different control points i we use. Henceforth, $\mathbf{b}_i$ will be used to denote the sum of the original biases.

Let $\mathbf{R}_C$ be the position vector of the control center. This would be a null vector if the control center were at the origin of our datum. To allow for a non-null vector here, define $\mathbf{r}_A = \mathbf{R}_A - \mathbf{R}_C$. Then, the range measurement to control point i is

$$C\Delta T = \sqrt{(\mathbf{r}_A + \mathbf{b}_i) \cdot (\mathbf{r}_A + \mathbf{b}_i) + (\rho_{Ai}) \cdot (\rho_{Ai})} \tag{2.19}$$

Note that the entire bias has been included on the control center-to-satellite vector. Part of it really belongs on the satellite-to-user vector, but it is easy to show that the end result will be the same to first order, whether it is put all on one or the other vector, or split between them. The effects are not separable.

By expanding (2.19), neglecting $\mathbf{b}_i \cdot \mathbf{b}_i$ compared to $\mathbf{r}_A \cdot \mathbf{r}_A$, and using $\rho_{Ai} = \chi_i - \mathbf{R}_A$ gives

$$(C\Delta T)^2 - (\mathbf{r}_A + \rho_{Ai}) \cdot (\mathbf{r}_A - \rho_{Ai}) = 2(\mathbf{r}_A + \rho_{Ai}) T_{\mathbf{b}i} \tag{2.20}$$

Because everything on the left-hand side of (2.20) involves measured or known nominal values, the left-hand side is considered known. The unknown bias $\mathbf{b}_i$ appears linearly on the right-hand side. Therefore, three copies of this equation, one for each of three distinct control points, allows for solution of the three unknown components in $\mathbf{b}_i$. This is true provided that the three-by-three matrix:

$$\begin{bmatrix} (\mathbf{r}_A + \rho_{A1}) \, T \\ (\mathbf{r}_A + \rho_{A2}) \, T \\ (\mathbf{r}_A + \rho_{A3}) \, T \end{bmatrix}$$

is nonsingular. The rows of this matrix are the transpositions of the vectors from the control center to each of the three control points. Nonsingularity requires that these three vectors not be coplanar. If the three control points are all sufficiently close to the control center as well as each other and the terrain is sufficiently flat, then an ill conditioned matrix will result, even if it is not exactly singular. Nevertheless, if the control points are not sufficiently close together, the assumption that the errors on the three satellite-to-user links are the same (i.e., biases) becomes harder to justify. There is a trade-off here that has not been fully studied yet. However, at least conceptually, it is possible to solve for the vector bias. It is the sum of the biases on the satellite position vector (i.e., satellite ephemeris bias errors) and biases on the downlink such as ionospheric delay (coverted to a position). The reader should note that the above solution process is also subject to error, as in the basic positioning problem.

Rather than trying to solve for the entire bias vector, another possibility exists. We could correct each individual round-trip distance measured to a user by an amount obtained from comparison of the measured value with a known control point and the computed value. It is not clear whether RDSS systems will generally use a vector or scalar version of the bias estimation process or some other method. Further analysis is needed concerning this question.

2.1.4.5 *The Three-Satellite Solution*

Comparison of the slopes in Figures 2.6 and 2.7 might give the impression that the sensitivity of user position error to altitude errors is higher than the sensitivity to ranging accuracy. The differences in units on these two curves contributes to this impression. It is true, however, that altitude errors are important, and thus it is of interest to see how accurately user position can be determined if a third satellite is available, instead of an altimeter or terrain map. A third equation like (2.9) and (2.10) thus replaces (2.13). An analysis exactly like the one in Section 2.1.4.2 has been carried out, and there are now nine input error contributors: three ranging errors $\Delta\rho_i$; three satellite radius errors ΔR_j; and three satellite angular errors $\Delta\theta_j$. The error sensitivity matrix $S(n)$ (or in local east, north, and up coordinates, $S'(n)$) is now three-by-nine. By placing the third satellite at 105° W, we obtained the results tabulated in Appendix 2.B for four different user latitudes in roughly the center of CONUS. For comparison with the two-satellite case of Section 2.1.4.4, all $\Delta\rho$ and ΔR values were set to 10 ft. The satellites angular errors were set to 0.4 μrad, and the user was again placed at 90° W and 30° N. The resultant RSS position error was 1309.5 ft, or about 403 m. This contrasts with the earlier two-satellite result of 63 ft.

The reason for the drop in accuracy is, of course, due to the poor GDOP that occurs here with all three satellites in the equatorial plane.

Table 2.7

Significant Error Contributors

Category	Subcategory or Contributing Source
Satellite errors	Radius and angles or radial plus normal components; Drift velocity uncertainty (of secondary importance).
Ranging errors	Receiver-related errors, i.e., jitter, thermal noise, *et cetera*; Ionospheric delay; Tropospheric refraction; Multipath effects; Timing (synchronization) between control-center and user equipment.
Third-dimension errors	Terrain mapping errors; Altimeter instrumentation errors; Combination of both (radar altimeter).
Calibration errors	Location errors in known control points; Location errors for control center; Spatial gradient effects, i.e., determine correction at control point, but use it at another point; Reference system conversions (i.e., satellite reference system to local coordinate grid); Barometric altimeter corrections.
User-motion models	Important if velocity is to be computed or in the case of time-recursive methods and in accounting for system delays.
Solution algorithm and numeric errors	Single-fix, multiple-fix algorithms; Computational finite word-length effects.

2.1.5 Comparison With System Architecture of Other Satellite Systems Capable of RDSS-Types of Services — Argos and Mobile Satellite Service

There are two other types of non-RDSS satellite systems that nevertheless are capable of some type of RDSS service. These are the French *Argos* location and data collection satellite system. and the mobile satellite service as implemented by the US-UK-controlled *Inmarsat* standard-C positioning service.

2.1.5.1 Argos

The Argos system offers capabilities for the satellite-based location of fixed and moving devices for the purpose of collecting environmental data. The system is a result of cooperative agreements between the Centre National d'Etudes Spatiales (CNES, France), the National Aeronautics and Space Administration (NASA, US), and the National Oceanic and Atmospheric Administration (NOAA, US). Pursuant to these agreements, CNES is responsible for constructing specialized electronics packages for integration into NOAA polar-orbiting meteorological satellites and subsequent launching by NASA aboard Atlas-Centaur rockets. Also, NOAA is responsible for separating Argos data from other data received from the weather satellite downlink, while CNES is responsible for the processing and distribution of the Argos data. Finally, CNES, NOAA, and NASA jointly agree on authorization of user access to the system via the purchase or manufacture of specialized user terminal devices called *platform transmitting terminals* (PTTs), which transmit at frequent intervals. Approximately every 100 minutes, a NOAA spacecraft passes overhead and receives, logs the time of, and records the PTT transmissions. When the NOAA spacecraft passes over an Argos telemetry station, the stored PTT transmissions are transmitted for data processing.

Theory of Operation

Location of PTTs is determined by calculation of the Doppler effect on received frequencies. Transmission frequency for all PTTs is fixed at 401.650 MHz. The satellite's received frequency at any instant can be used to define the field of possible positions for the PTT. The field is in the form of a half-cone, with the satellite at its apex, the satellite velocity vector as axis of symmetry, and the apex half-angle (A), such that

$$\cos (A) = [(f_r - f_e)/f_e] \times (c/V)$$

where

c = speed of light;
V = satellite speed relative to PTT;
f_e = transmission frequency = 401.650 MHz;
f_r = received frequency.

As shown in Figure 2.11, one location cone is obtained for each Doppler measurement. The altitude of the PTT forms part of a sphere, the *altitude sphere*, which is also known. The intersection of the various location cones with the altitude sphere thus gives two possible positions of the PTT, which

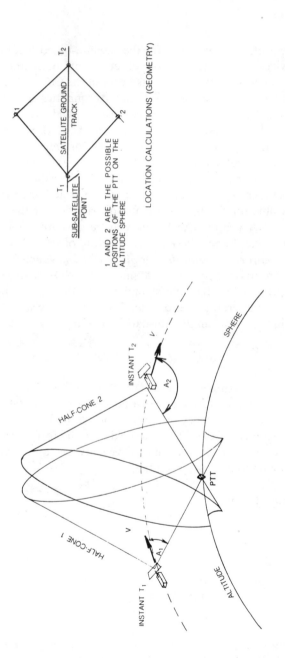

Figure 2.11 Polar Orbit Radiodetermination
Source: Service Argos, Inc.

are symmetrical with respect to the satellite ground track. To find which of the two positions is correct, additional information is required, such as previous locations or range of possible speeds.

Other information required for location determination by using Doppler measurement techniques includes satellite orbital parameters and precise time-coding of measurements. Satellite oribtal parameters are obtained by 11 orbitography PTTs. These PTTs are at accurately known geodetic locations. They are equipped with very stable oscillator yielding orbital parameters on a daily basis and permitting extrapolation onto the next day's orbit. The system currently permits determination of satellite position to within 30 m in the ground-track direction and 250 m in the cross-track direction A *time-coding* PTT, featuring high-precision time coding and extreme frequency stability (cesium clock), is located in Toulouse, France. This PTT is used to monitor the stability of the onboard oscillator and to align all measurements with the same time scale (GMT) with a mean precision of 12 μs.

Before calculations proper are performed, certain geometric tests are carried out to eliminate any PTTs for which there is no guarantee of acceptable accuracy. On average, 33% of PTTs received during a satellite pass are eliminated by this test. The main causes of rejection are (a) excessive frequency deviation (more than 24 Hz difference in two passes for the same PTT), (b) unsatisfactory convergence (oscillator instability during a pass of more than 4.1^{-5}, and (c) unacceptable distance from ground track (i.e., positions determined for PTTs that are either too close to, or too far from, the ground track are usually inaccurate).

Figure 2.12 shows the mean number of locations per day as a function of PTT latitude that are possible with the Argos system. Location accuracy is largely a function of PTT oscillator stability and platform movement, as indicated in the figure. Oscillator stability is important in a short, medium, and long term. The short term, defined as up to 100 ms, concerns the duration of message transmission. Short-term instability leads to inaccuracy in the calculation of the Doppler effect and results in random error for location calculations. If short-term stability is worse than 10^{-8}, the message is ignored and neither data collection nor location are performed by the satellite. The medium term, defined as 20 minutes, relates to the duration of PTT visibility. Medium-term instability results in the frequency drift during the satellite pass, thus leading to inaccurate calculation location. The main cause here is temperature variation during the satellite pass. In general, the necessary stability can be readily achieved by placing a standard oscillator in a suitably insulated housing (made of polystyrene, for example). Argos requirements call for a minimum medium-term stability of 2.1^{-7}, corresponding to a location within several tens of kilometers. Long-term is defined as the period separating two satellite passes over a given PTT, or about 100 minutes. Long-term instability

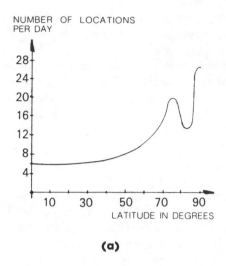

(a)

MEDIUM TERM STABILITY		ACCURACY IN METERS	
%	in Hz/mn	in 65 % of cases	in approx. 95% of cases
2. 10^{-9}	0.04	150	500
5. 10^{-9}	0.10	500	1300
10^{-8}	0.20	1100	2000
2. 10^{-8}	0.40	2100	3600
5. 10^{-8}	1.00	4000	6000
10^{-7}	2.00	5500	9000
2. 10^{-7}	4.00	approx. 50 km	
> 2. 10^{-7}		calculation aborts	

(b)

Figure 2.12 Argos Service Characteristics:
(a) Mean Number of Locations per Day
(b) Location Accuracy
Source: Service Argos, Inc.

is compensated in the Argos system if it does not exceed 10^{-6}. The PTT is considered to be stationary at the moment of transmission. Therefore, any movement can result in error; thus, a speed of 1 m/s leads to a location error on the order of 200–300 m.

Satellite Orbits and Communication Parameters

One of the most significant differences between the Argos system and RDSS systems is satellite orbit. In contradistinction to the geostationary orbital characteristics of an RDSS system, the Argos system has the following orbital characteristics:

- Circular orbit;
- Altitude of 830 km ±18 km and 870 km ±18 km for each of NOAA/Tiros-N satellites carrying an Argos payload;
- Polar orbit—the satellites see both the North and South Poles once each orbit. Inclination (angle between the equatorial and orbital planes) is 98.7°;
- Sun-synchronous orbit—this means that the orbital plane rotates about the polar axis at the same speed as the earth rotates about the sun; that is, one complete rotation per year. Each orbit therefore transects the equatorial plane at fixed local solar times. From the user's point of view, this means that a given PTT comes into satellite visibility at the same local solar time every day;
- Period—it takes approximately 101 minutes to complete one orbit;
- Number of orbits per day—approximately 14 for each satellite.

This set of orbital characteristics guarantees complete coverage of the earth's surface. The orbital planes of the two satellites are mutually offset by 75°. Orbital altitudes are also different so that the period of one satellite's orbit is approximately one minute longer than that of the other. This ensures that a given point on the earth is not seen at the same instant by the two satellites.

The Argos PTTs must conform to a specified communication format, as indicated in Figure 2.13, which is modulated by using split phase-pulse modulation. In particular, each transmission has the following sequence:

- 160 ms of unmodulated carrier to allow the Argos onboard receiver to lock onto the carrier;
- A 15-bit preamble to synchronize the Argos onboard equipment with the message bit rate;
- An eight-bit format synchronization word, followed by one spare bit and four bits defining sensor data length (the number of 32-bit blocks);
- The PTT ID code, assigned by Argos and encoded in 14 bits;
- Six error check bits;
- 32 to 256 bits of sensor data in steps of 32.

A wide variety of Argos PTTs are available that meet the above-described certification standards of the Argos organization. Most PTTs are in maritime applications, such as floating buoys, moored buoys, and shipboard platforms.

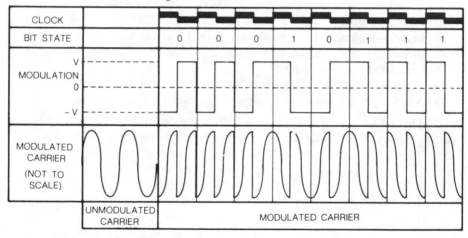

UNMODULATED CARRIER LENGTH T1	MODULATED CARRIER : LENGTH T2					
	PREAMBLE	FORMAT SYNC.	INITIALIZATION	NUMBER OF 32-BIT GROUPS	ID N° (+ Check bits)	SENSOR DATA
T1 = 160 ms ± 2.5 ms	15 bits (= 1)	8 bits (00010111)	1 bit (= 1)	4 bits	20 bits	N × 32 bits (1 < N < 8)

PTT MESSAGE STRUCTURE AND FORMAT

(a)

T_B = BIT PERIOD

CLOCK								
BIT STATE		0	0	0	1	0	1	1	1

MODULATION   V / 0 / - V

MODULATED CARRIER (NOT TO SCALE)

UNMODULATED CARRIER	MODULATED CARRIER

(b)

Figure 2.13 Polar Orbit Radiodetermination Protocol:
(a) Message Structure
(b) Format
Source: Service Argos, Inc.

Less common are land PTTs such as remote meteorological, snow data, and hydrological stations. Fairly exotic PTTs, which must meet the communication link stability specifications under very extreme conditions, include polar ice bouys, atmospheric balloon stations, and animal tracking platforms. A fascinating example of the utility of radiodetermination via satellite is the effort being made to study and protect certain animal species with information about their distribution and migration. A number of satellite-based programs are underway to track dolphins, basking sharks, leather-back turtles, whales, birds, and wild pigs. Typical technical problems posed by applying Argos technology to animal tracking are given below:

- Short duration of satellite visibility and unsatisfactory propagation characteristics;
- Difficulties in developing PTTs that can withstand the high pressures and mechanical forces encountered in tracking marine animals;
- Relatively short battery lifetimes (although this can be increased when tracking marine species by fitting a switch that turns off the PTT when the subject is diving, and by using solar cells when tracking land animals);
- The weight of the PTT, which must be negligible in relation to the animal's own weight.

2.1.5.2 Mobile Satellite Service and Inmarsat Standard-C

The mobile satellite service is a communication capability wherein bandwidth is divided by frequency into pairs of channels, which are, in turn, allocated in pairs to mobile communicators. The frequency division is accomplished to optimize the system for voice communication. This normally implies channel bandwidth of anywhere from 5 kHz to 50 kHz, depending on various practical considerations. If these channels are also used to derive positioning information, then the mobile satellite service is also providing a radiodetermination capability. If the channels are simply used to relay positioning information determined in another manner (human keying from Loran-C, *et cetera*), then the mobile satellite service is merely providing a position reporting, not a radiodetermination capability.

Different mobile satellite services have been contemplated for maritime, aeronautical, and land-mobile applications. The Inmarsat system is the standard maritime mobile satellite service. Baseline land-mobile and aeronautical mobile satellite services have been described in CCIR Reports 770-1 and 559-1, respectively.

Theory of Positioning Technique

Narrow-band ranging, the only type of range acquisition possible with mobile satellite systems, is accomplished via continuous audio-frequency modulation. This technique, also called *tone ranging*, may be accomplished in either analog or digital form. Narrow-band ranging is significantly less precise than wideband ranging methods employed in RDSS systems, but experiments have demonstrated accuracy on the order of 0.1 mile.

With the analog approach, distance is measured by means of phase-comparison techniques in which a transmission consists of a single modulated by one or more sinusoidal audio frequencies. The signal is retransmitted by repeaters located in the vehicle and at the satellite, and the total round-trip propagation time delay observed at the originating station is determined by measuring the phase of the received signal with respect to the transmitted one. If the one-way distance is r, then the difference in phase (ϕ) between the transmitted and received signals is

$$\phi = 4\pi f_m / c \, (\text{rad})$$

where

f_m = modulating audio frequency;
c = speed of radio-wave propagation.

For a given phase-difference measurement error, the corresponding distance error is inversely proportional to the audio frequency. To minimize distance errors it is therefore desirable to use as high a ranging audio frequency as possible. The relationship between the distance measurement precision (∂_r) due to noise and the postdetection audio-frequency signal-to-noise (S/N) ratio is (disregarding integration improvement):

$$\partial_r = c / [4\pi f_m (2S/N)^{1/2}]$$

A factor affecting the choice of audio frequencies is the *ambiguity distance*. A phase measurement can only be made over the interval of 360°. Therefore, there is an ambiguity in the number of full cycles of phase delay, and this phase measurement ambiguity corresponds to an ambiguity (r_{amb}) in the distance measurement of

$$r_{\text{amb}} = c / 2f_m$$

The ambiguity can be resolved by coding the signals, or sending additional audio frequencies.

In general, continuous audio-frequency modulation does not provide the same type of immunity to the effects of multipath as can be attained with wideband PN ranging because it becomes difficult to separate in time the direct signals from the multipath-reflected signals. In some cases, however, it may be possible, by using filters with narrow passbands, to filter the signal waveform to average the multipath error.

A *digital* tone-ranging technique has been developed in Japan to obtain measured ranging data via satellite. In this digital tone-ranging technique, the rectangular waveform is used as a measurement signal, and so this method differs from the analog techniques described above, where we use the sinusoidal waveform. Measurement principle, estimated measurement error, necessity of ambiguity check, and multipath effect are all similar to those of the analog methods. However, in addition, this technique characteristically inserts low-frequency signals into the precision measurement signal so as to remove ambiguity.

The measurement signal, with a sufficiently high frequency for obtaining the necessary measurement accuracy, is shaped into a rectangular waveform. A small part of the rectangular waveform is replaced with a coded signal, synchronized with the low-frequency, ambiguity-removal signal. The carrier signal, now phase modulated by this digital waveform, is transmitted and received via satellite. As is customary, the range is obtained by measuring the round-trip delay of the code extracted from the digital waveform.

By utilizing such a digital waveform in a range measurement system, the number of low-frequency signals necessary for ambiguity checking can be made smaller than that of the analog techniques. This is because the zero-crossing of the low-frequency signal is determined by detecting the inserted code with the same bit rate as the rectangular high-frequency signal. In summary, digital tone ranging is as amenable to narrow-band systems as are analog techniques, but with fewer drawbacks.

The *tone-code ranging* experiments, using the NASA ATS-6 satellite, provide a good example of how to apply the theoretical concepts. A mobile unit was tracked via lines of position derived from active tone-code ranging through the very high gain ATS-6 satellite and by passive reception of accurate timing signals from a geostationary weather satellite known as GOES. Tone-coding ranging interrogations were transmitted from a control center, via the ATS-6 satellite, alternatively to both the mobile unit and a fixed based station, of which true coordinates were well known. The position fixes of the base station derived by tone-code ranging were compared in a computer with its true position to obtain differential corrections for positioning errors. The mobile unit received accurate clock signals from the GOES satellite at one-second intervals, then fed these clock "ticks" into the "playback" response to the tone-code ranging interrogations from ATS-6.

The tone-code ranging signals that produced the ATS-6 lines of position consisted of 1024 cycles of a 2.4414 kHz tone, which was frequency modulated on the 1650 MHz band carrier with a deviation of 5 kHz so that the signal was limited to a 15 kHz bandwidth. The mobile unit's radio equipment matched self-generated 2.4414 kHz tones to the received tone's phase by averaging over 256 cycles of the received tone to determine its phase. When the code received from the satellite matched the code that was hard-wired into the mobile unit's correlator, an output was generated to signify the time of arrival of the satellite signal. This output stopped the counter that was counting time since the last one-second GOES tick. The output then generated a transmission to the satellite, consisting of the tone burst, unique ID code, internal time delay measurement, and the count in the GOES interval counter. Concepts very similar to this underlie all radiodetermination techniques in the mobile satellite service.

Because the above-mentioned narrow-band ranging techniques are generally less accurate than other wideband ranging systems available, it is likely that most positioning information associated with mobile satellite systems will simply be data relay. For example, a mobile satellite service terminal may automatically receive the output of an RDSS transceiver or a radio navigation (satellite or terrestrial) receiver. Indeed, this is customary practice today in the Inmarsat system, although the positioning data are often entered manually and hence are prone to error. The important point to remember is that using a communication channel to simply *relay* positioning information does *not* convert the channel into a radiodetermination channel. A radio communication channel must itself be used to *derive* positioning information in order to make use of the frequency allocations and characteristics associated with radiodetermination.

The question may therefore arise as to whether systems such as Inmarsat, land-mobile satellite, and aeronautical mobile satellite become radiodetermination satellite systems if they indeed derive positioning information from their channels by using tone-code or similar ranging techniques. The official bodies that have considered this question repeatedly answer it in the negative. After exhaustively reviewing the issue, the FCC, for example, noted that if one's primary purpose is radiodetermination, then the system design objective is to maximize positioning accuracy and capacity. This is accomplished with wideband, randomly accessed, time-shared channels. Hence, systems with these characteristics (Geostar, Mobile Communications Corporation of America, McCaw Communications) were entitled to avail themselves of the RDSS frequency allocation. The fact that such systems also provided a simple mobile communication capability did not convert them into mobile satellite systems because it was clear that such was not their principal design objective.

Nonetheless, if one's primary purpose is mobile voice communication, then the system design objective is to maximize the number of acceptable quality voice-band channels. This is accomplished with narrow-band, frequency channelized, demand-assigned channels. Systems with these characteristics (Omninet, Mobilesat, Skylink) in the United States were not allowed to use the RDSS allocation, but were considered to be mobile satellite service license applications. The fact that such systems also provide a simple tone-code ranging capability did not convert them into RDSS systems because it was clear that this was not their principal design objective.

2.2 SPACE-SEGMENT DESIGN

Space-segment design considerations for RDSS require providing as high a satellite G/T as possible for the inbound link, adequate EIRP for the outbound link, adequate provision for the trunking links to the control center, and overall minimization of complexity and potential failure modes. As with all space-segment design issues, practical problems dominate. For RDSS these problems will often appear as constraints on the implementation of an RDSS payload aboard a host spacecraft that may be optimized with other purposes in mind.

Figure 2.14 is a block diagram of an eight-beam RDSS payload system developed by COMSAT as a dedicated spacecraft for Geostar Corporation and submitted by the US government for advance publication to the ITU as USRDSS East, Central, and West. The payload fully meets the reference RDSS link budget provided in Section 2.1.2. It is a particularly complex payload by virtue of carrying a PN generator on-board from which to generate spread-spectrum signals. Figure 2.15 is a block diagram of a single-beam RDSS package (portions within dashed boxes) designed by Aerospatiale for Locstar Corporation as a payload to be added onto a C/S-band spacecraft. It is an especially simple payload package, partly because the satellite already enjoyed an S-band subsystem (for direct television broadcasting).

2.2.1 L-Band Receiving Subsystem

The fundamental requirement of the L-band system is to receive user transmissions and route them into C-band transponders for retransmission to the control center. Physical constraints may either force or preclude use of a common antenna assembly for L and S band. The frequency difference between these two bands makes such a common antenna difficult, but not impossible.

Planar arrays are an attractive choice for RDSS antennas. Table 2.8 provides some basic parameters for a Ball Aerospace L-band planar array, considered for Figure 2.15, which is capable of being mounted on most geostationary spacecraft.

In order to meet the CCIR reference RDSS link budget, a significantly larger antenna must be used in the receiving subsystem. EOC gain of 29.0 dB would bring G/T up to 1.0 dB, the CCIR baselined figure. To do so at L band, however, requires a 15-ft-diameter parabolic reflector.

A simplified block diagram of a generic L-band RDSS receiving subsystem is given in Figure 2.16. We should recall that every RDSS spacecraft will normally include a receiver subsystem because each constitutes either a primary or back-up ranging path. Hence, three receiving subsystems are needed for the two-range RDSS method and four are needed for the three-range

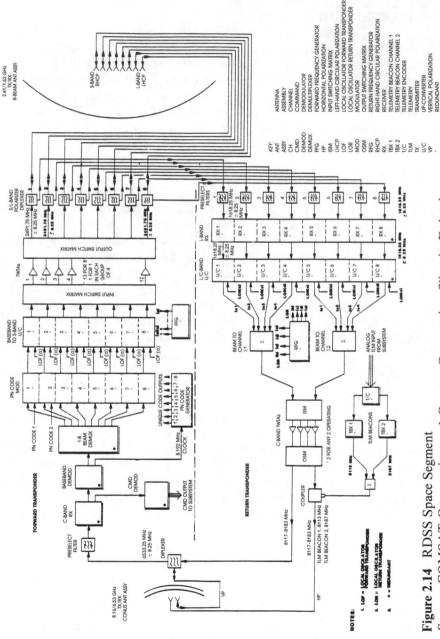

Figure 2.14 RDSS Space Segment
Source: COMSAT Corporation and Geostar Corporation filing in Federal Communications Commission, General Docket No. 84-689 (1985).

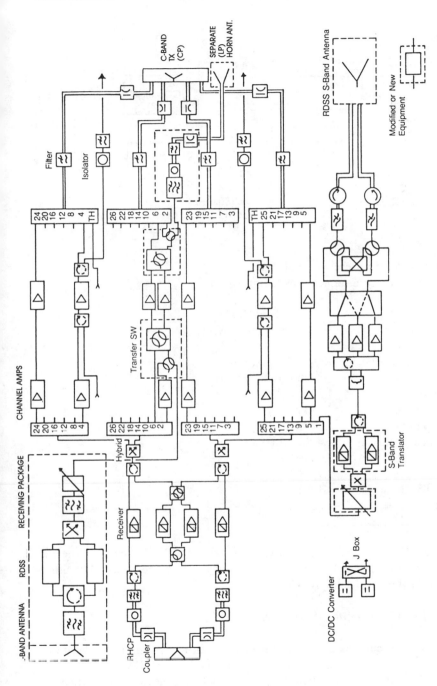

Figure 2.15 RDSS Add-On Payload
Source: Locstar Corporation and Geostar Corporation, 1986.

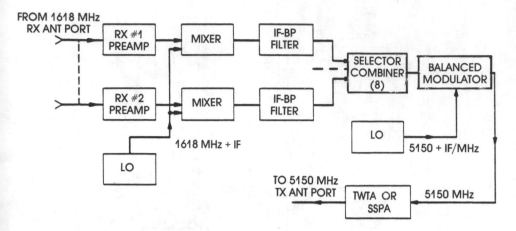

Figure 2.16 RDSS Receiving Subsystem
Source: Federal Communications Commission

Table 2.8
L-Band Planar Array

Characteristic	Parameter
Frequency	1610.0 – 1626.5 MHz
Size	1.321 × 0.724 m
Mass	3.58 kg
Polarization	Left-hand circular
Axial ratio	1.4 dB
EOC gain	23.0 dB
Coverage	3.7° × 6.7°
VSWR	<1.5:1
G/T	–5.0

method. Although the S-band subsystem is critical to RDSS performance, it need not be implemented more than twice (one primary one back-up) in either the two- or three-range approach.

2.2.2 S-Band Transmitting Subsystem

For a difficult case, consider the eight-beam RDSS spacecraft with on-board PN code-generating capability (Figure 2.14). The 6 GHz received signal, separated from the transmitted frequency signal by means of a frequency-selective diplexer, is passed through a preselector filter used to guard against out-of-band interference. A C-band receiver with a low noise front-end to achieve the required noise figure and hence G/T, down-converts the C-band signal to a convenient intermediate frequency (IF) with further amplification where the command signal is separated from the receiver.

The IF signal is then applied to a baseband demodulator, which recovers the ground-originated baseband signal format containing time pulses, user address, beam-number information, and interrogation messages for retransmission to the users. The baseband information is, in turn, passed to a beam demultiplexer (*demux*), the function of which is to route the appropriate digital information stream to the require beam.

The eight outlets of the demultiplexer are connected to eight *pseudorandom code modulators*, which impose unique PN codes on the information stream for each beam, providing the beam-to-beam isolation required for overlapping satellite coverage. The PN code generator is locked to the recovered synchronizing clock signal. The PN code modulators directly feed the eight independent up-converters which up convert the beam directed messages, to S band, with the required filtering so as to limit the signal spectrum in each beam to 16.5 MHz.

The S-band up-converter assembly, in turn, feeds the eight active traveling wave tube amplifiers (TWTAs), each delivering approximately 100 W of RF power. The TWTAs are preceded and followed by an appropriate switching matrix to achieve the required redundancy and power amplifier reliability. The active TWTA outputs are routed to the appropriate diplexers-polarizers and then to the antenna feed horns.

2.3 CONTROL-SEGMENT DESIGN

The control-segment of an RDSS system consists of two major elements — one radio frequency (RF) and the other data processing (DP). Each is needed for both inbound and outbound services. While not facile, the DP portion of the control center is similar to other real-time, data-base-dependent computer networks. The RF part of the RDSS control segment is, however, rather unique.

Figure 2.17 is a functional block diagram of a RDSS RF control center, or, in vernacular, "hub." The hub is characterized by much parallel activity. Its highest level function is to command a master clock for the systems that drive the outbound chip rate, referenced by the CCIR at 8.192 MHz, or half the 16 MHz bandwidth. Figure 2.18 offers a high-level description of the data network that must bridge the control center's RF and DP activities.

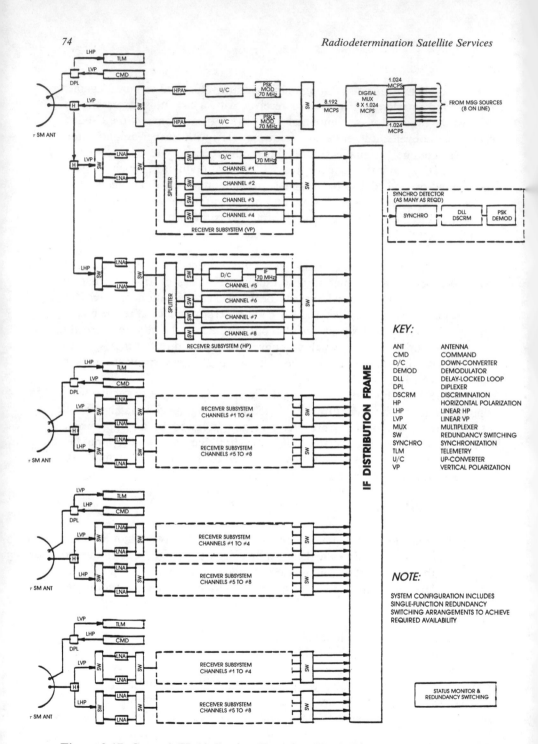

Figure 2.17 Central (Hub) Station Function Block Diagram
Source: Federal Communications Commission, General Docket No. 84-689 (1985).

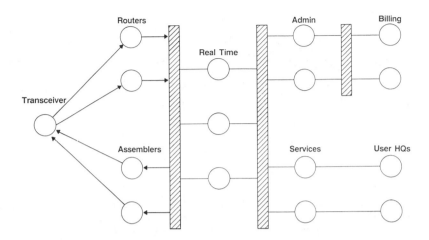

Figure 2.18 RDSS Data Network

2.3.1 Inbound Traffic Hardware and Software

On the receiving side, the central hub monitors the downlink transmissions from the satellites, and seeks, detects, acquires, and tracks all incoming transmissions. All of this effort is going on simultaneously (see Figure 2.2). Once synchronization is obtained, the data portions of the transmissions must be demodulated and decoded, and synchronization includes precise tracking of the incoming PN sequences with a phase measurement made, allowing measurement of each user's round-trip time to a small fraction of a chip. The decoded data and precise time measurements are "handed off" to the DP segment for further processing.

The inbound links would normally employ a frame structure, as illustrated in Figure 2.19. Users transmit by spread-spectrum time-division multiplexing (TDM) in short bursts. Each response includes a synchronization preamble, an acquisition segment ending with a unique word, and message bits. When the reply signals from an RDSS user are received at the control center, it must be determined exactly where in the superframe the user transmission originated in order to make the appropriate range calculations. This is done by including within the inbound transmission the outbound frame number corresponding to the time that the user initiated the transmission.

2.3.2 Outbound Traffic Hardware and Software

The data upon which the outbound chip rate is imposed may be conceptualized into structures, such as frames comprising certain numbers of chips and superframes comprising some number of frames. Each frame should have

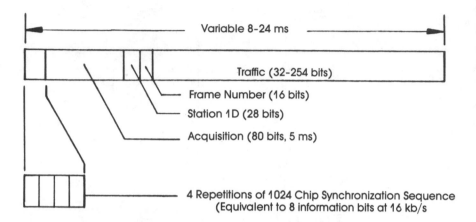

Figure 2.19 Typical Format for Inbound Transmissions: Minimum
Information Bit Rate 16 kb/s
Source: Federal Communications Commission, General Docket
No. 84-690 (1985).

a defined protocol, such as standard HDLC. A typical frame format is shown
in Figure 2.20.

   Each frame is time logged, has a frame number, contains data, leads with
spread-spectrum acquisition aids, and serves as a time epoch t_0 for replies from
the user's transceivers. The CCIR reference RDSS system transmits the
outbound data at the 64 kb/s information rate derived from a 128 kb/s data
rate, with one-half-rate FEC. Clearly, we can see that there is considerable
latitude in frame structure design.

2.3.3 System Ephemeris

   We should note that the control segment also normally provides standard
tracking, telemetry, and control services for the RDSS spacecraft. Recall from
Section 2.1.3 that, from the standpoint of the control center, the satellite
coordinates $(r_{Si}, \theta_{Si}, \phi_{Si})$ may be treated as fixed constraints because they are
associated with geostationary satellites. While this is approximately true, minor
perturbations will normally occur in the orbital position of such a satellite due
to the gravitational influences of the sun and moon. For this reason, the satellite
coordinates are preferably left as variables and inserted by the control center's
computer during the course of each user position calculation. The coordinates
thus inserted may then be continually updated, based on known satellite
position schedules stored in the computer memory, or periodic direct measurements of the satellite positions.

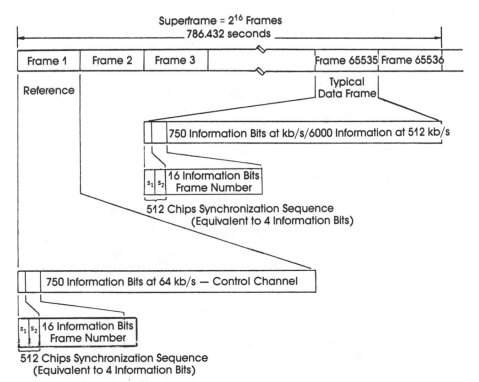

Figure 2.20 Outbound Superframe Structure
Source: Federal Communications Commission, General Docket No. 84-690 (1985).

2.3.4 Control-Center Data Processing Hardware Configuration

A distributed hardware configuration makes the most sense for an RDSS control center. A functional diagram of baseline data processing hardware is provided in Figure 2.21. This configuration should be considered in connection with the entire control-center data network depicted for a baseline case in Figure 2.18. What results is essentially a local area network consisting of the following generic processing nodes: routing, real-time activity, administrative, store and retrieve, services, billing, subscriber activity, and assembly. In Figure 2.21 the administrative processor is handling the store and retrieve and the services processes. All machines depicted in the figures may be implemented quite reasonably in practice with networks of "midicomputers," such as DEC Microvax or HP Spectrum for a throughput capacity of up to several hundred thousand transactions per hour.

The processors in the local network communicate with each other through a bus, such as Ethernet 802.3 connections. Redundant front-end (prereal-time processing) and back-end (postreal-time processing) buses are

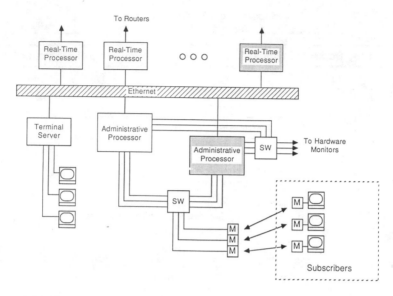

Figure 2.21 Central Hardware Configuration

recommended to ensure continuity of real-time data processing. The configuration of the critical generic processor types is described below.

2.3.4.1 Router and Assembler

The main function of the *router* (see Figure 2.18) is to receive transceiver data from the demodulators and pass that data to the appropriate real-time processor. The purpose of the *assembler* is to transfer transceiver data to the modulators. In addition, the router and assembler communicate with the administrative processor for reporting performance data and receiving control-center network information. Each router and assembler should be implemented as a pair of primary and alternate processors. Additional pairs of routing and assembling processors will be needed to accommodate higher levels of system utilization.

2.3.4.2 Real-Time Processor

The function of the *real-time processor* is to receive transceiver data from the router, perform required real-time processing, and transmit data to the assemblers for outbound transmission. Each real-time processor should have sufficient memory for storing all required processing tables and buffers as well as sufficient disk storage for its operating system, application software modules, and data files necessary for starting real-time operations. Multiple primary and alternate real-time processors are recommended for system reliability.

Data that are stored at the real-time processor and required for position determination include benchmark locations, satellite positions, and a topological data file. Other data stored at this process include a keyboard message file, transceiver location files, and atlases.

2.3.4.3 Administrative and Subscriber Processors

The purpose of the *administrative processors* is to provide a means of operating, controlling, and monitoring the control center's data processing system. Operation includes an interface to all processors for managing the local area network. Primary and alternate administrative processors, sharing a common pair of disks to ensure identical data bases, are recommended for redundancy. Data stored in the administrative processor include subscriber profiles, an event log, a system status table, and keyboard message files.

Special functions such as billing, transceiver services, and store and retrieve may also be accomplished by the administrative processor, or they may be spun off to separate processors. The *subscriber processor* provides access to RDSS transceiver data by system users accessing the control center from fixed sites over normal data lines. The subscriber processor communicates with the administrative processor for validation of transactions and obtaining transceiver state data.

2.3.5 Control-Center Software Structure

The overall control-center software structure for generalized RDSS service is shown in Figure 2.22. This structure provides a high-level description of the information flow handling that must be accomplished as depicted by using data flow diagrams. These high-level data flow diagrams must be developed into sets of lower level, but significantly more detailed, diagrams. At the lowest levels rigorous definitions must be applied so as to write actual programming code.

The basic components are administrative, real-time, services, and subscriber access. This structure, which parallels the previously described hardware considerations, results in the major data bases residing on those processors that use them. In addition, the data flows between processes are buffers of information that pass as large messages over the 802.3 buses. This data flow handling process results in an efficient use of processor and system resources.

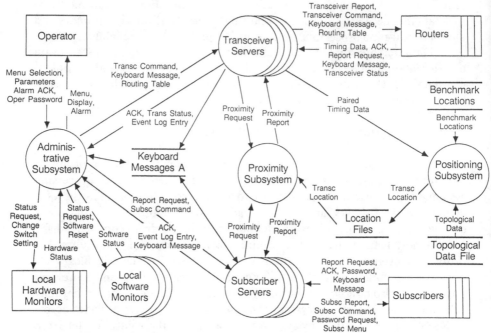

Figure 2.22 Central Software
Source: Geostar Corporation, "International Gateway Station," Contribution
to ITU CITEL Seminar on MOB-WARC-87.

2.3.5.1 *Administrative Process*

A detailed data flow diagram for the *administrative subsystem* is provided
in Figure 2.23. The basic purpose of an administrative subsystem is to have
centralized responsibility for network monitor and control functions.

All other processes would report performance data to the administrative
processor for storage and subsequent retrieval. For example, note from either
Figure 2.22 or 2.23 that transceiver and subscriber service data are sent into
the administrative subsystem (via the subsystem's communication process in
Figure 2.22). Such data may consist of acknowledgements (ACK), logged
events, messages, and diagnostic telemetry on transceivers. By retrieving all
or part of these data from memory, an administrative process operator may
command the billing and accounting routine to generate periodic invoices.

We should note here that all transceivers in an RDSS system, no matter
how varied and dispersed, ought to be considered part of a highly distributed
RF and DP system. For example, note how keyboard messages and transceiver

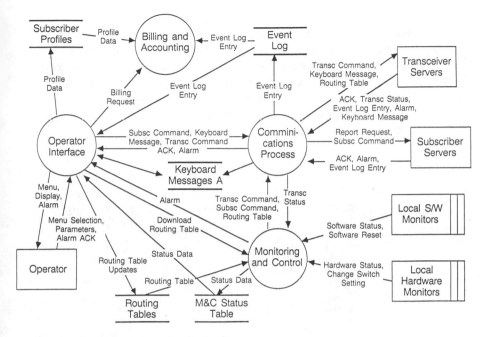

Figure 2.23 Administrative Subsystem
Source: Geostar Corporation, "International Gateway Station," Contribution to ITU CITEL Seminar on MOB-WARC-87.

status data are communicated from the transceiver server process to the administrative subsystem's communication process. The transceiver status data, such as a low battery power bit or other piece of telemetered hardware data, is referred to the monitoring and control process. Alarms might be sent to the operator, from which either a human or machine contact with the subscriber might be initiated, or switch settings might be altered to change the alarm threshold. Messages may be stored long enough to obtain subscriber data on the intended recipient, and then retransmitted via the communication process.

We may also observe from Figures 2.22 and 2.23 that the administrative subsystem passes key data from tables to all processors. Such data is used for routing interprocessor and intraprocessor messages within the local area network. Every node in the system communicates directly with the administrative subsystem to receive key information that each requires in order to perform its processing function.

2.3.5.2 Real-Time Process

The *real-time process* bridges the transceiver server and the router for the inbound link, and the subscriber server and the assembler for the outbound link. Figure 2.24 provides a data flow diagram of the inbound link activity from the transceiver server's perspective. Outbound link activity is similar. In general, the functions of the real-time process are to receive transceiver data, determine the processing steps that must be performed on each packet, pass packets to the services process, and transmit transceiver data.

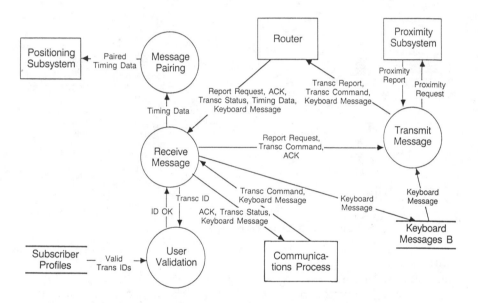

Figure 2.24 Transceiver Server

Note that a wide range of data arrives at the receive-message routine (see Figure 2.24) from the router. Operations performed on these data include spinning it off to the positioning subsystem, validating the transceiver ID code via data-base check with the subscriber profile, generating a response such as an acknowledgement, or complete RDSS position determination and message switching operation. Of course, many additional operations are possible for specialized systems. The key ingredients, however, are delineated step by step in Figure 2.25. These are to generate an immediate acknowledgement and determine transceiver ID validity, determine positioning and proximity, and retrieve dynamically stored keyboard messages. These three ingredients are then combined into a formatted transmission and queued for outbound spread-spectrum coded transmission.

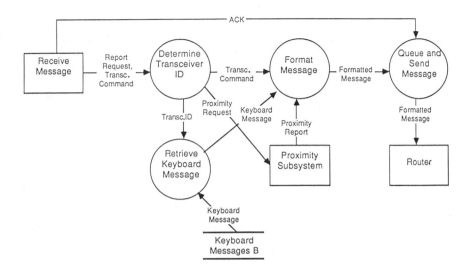

Figure 2.25 Transmit Message

2.3.5.3 Services Process

The *services process* performs altitude determinations, iterative position calculations, and proximity references ("look-ups") as required for incoming transceiver messages. Figure 2.26 provides a software overview of the positioning subsystem function. The paired timing data arise from multiple responses being received by the control center at slightly different times. This is due to a single transceiver response arriving via each of two or more satellite ranging and relay paths. Responses from benchmark transceivers at known locations permit continuous recalibration of satellite empheris, which enter a table from which is drawn the position calculation algorithm. A flow chart illustration of the sequence of operations bearing on the position determination is provided in Figures 2.27–2.29. Below we present a detailed examination of this sequence of operations as a general case example.

Let us first refer to Figure 2.27, wherein an RDSS program may commence at the START block and await the occurrence of the first signal TP_0, which happens concurrently with the transmission of the first interrogation signal by the control center. When the TP_0 signal appears, the program immediately proceeds to read the difference value ΔTP_0, and the current clock pulse time T, derived from the highly accurate local clock. The program next proceeds to block 330, where the interrogation signal transmission time t_0 is computed by arithmetically combining the difference value ΔTP_0 with the digital clock pulse time T_c. In block 332 the computed value of the interrogation signal transmission time t_0 is stored in one of a number of recirculating memory

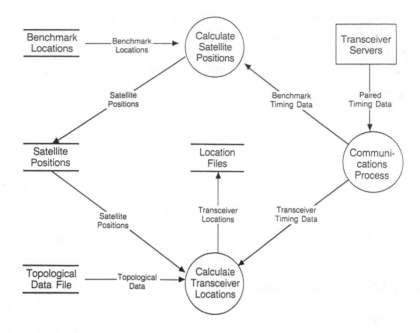

Figure 2.26 Positioning Subsystem

locations so the it will be available for use in subsequent positioning calculations. The steps represented in blocks 326–332 are obviously required only
at the beginning of a sequence of interrogation signals. Recall that because
the orbital height of the RDSS satellites is roughly 35,000 km, it can be
demonstrated that no return signals from transceivers (*automatic beacons
transponders*, or ABTs) can be expected to arrive at the control center until
approximately 0.47s after the interrogation signal is transmitted from the
control center.

Blocks 334–357 describe time-measurement integrity steps that must be
taken in any RDSS system. The particular steps described in these blocks are
for the three-range method patented by Dr. Gerard K. O'Neill. Note that
decision block 350 (Figure 2.28) refers to a time window, 0.04 s, which
represents the maximum difference in transit time for the transceiver response
relayed via the closest and farthest points on the earth from a given spacecraft.

With acceptable time measurements, the program proceeds to block 358
(Figure 2.29), where it forms the differences (t_2-t_0), (t_1-t_0), and (t_3-t_0). Next,
in block 359, the differences are inserted into the position-computation equations described in Section 2.1.3 to solve for the possible positions of the
identified user in terms of spherical coordinates (r_0, θ, ϕ). As we noted

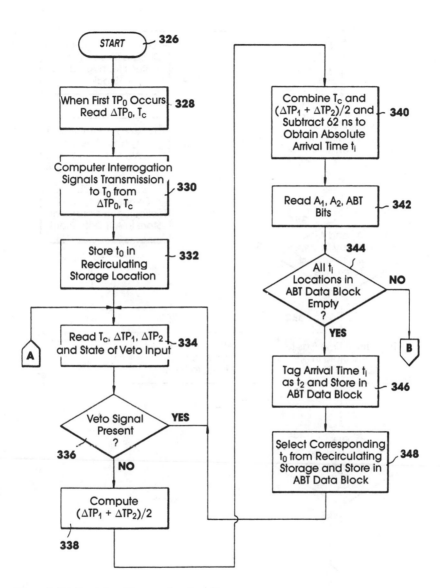

Figure 2.27 Position Determination I
Source: Courtesy of Dr. Gerard K. O'Neill

previously, it is a consequence of the employed equatorial satellite pattern that two different solutions to the position equations occur for each valid set of time differences (t_2-t_0), (t_1-t_0), and (t_3-t_0). One solution is the true position of

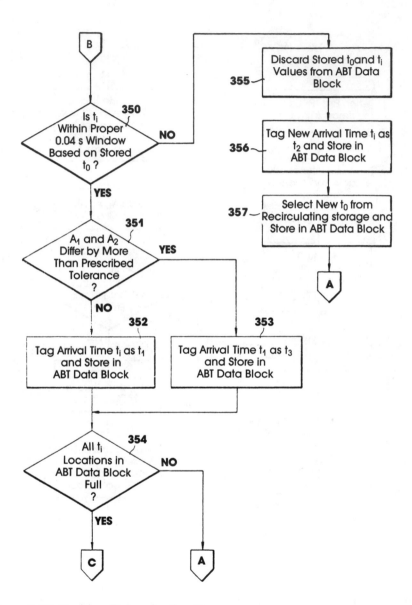

Figure 2.28 Position Determination II
Source: Courtesy of Dr. Gerard K. O'Neill

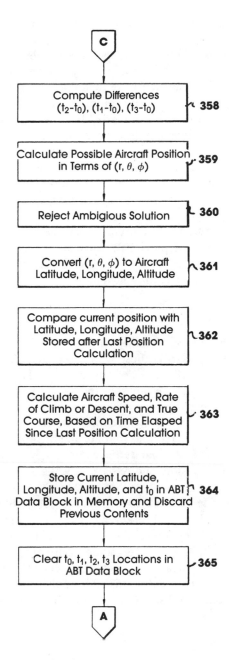

Figure 2.29 Position Determination III
Source: Courtesy of Dr. Gerard K. O'Neill

the vehicle, and the other is an ambiguous solution corresponding to its mirror-image position with respect to the equatorial plane of the earth.

The ambiguous solution is rejected in block 360. This will ordinarily be simple because most systems operate in only one hemisphere. However, it is possible to design the system software such that, as the latitude of an aircraft approaches zero, the ground station computer calculates the sign and approximate magnitude of the time derivative $d\theta/dt$, based on one or more previously stored position calculations for that aircraft and the time elasped since their computation. When the latitude of the aircraft subsequently reaches zero, the results of this calculation may be used to determine whether the aircraft is in fact crossing from the northern to southern hemisphere, or *vice versa*.

Now that we have isolated the true location, the computer next proceeds to block 361 of the program, where the aircraft position in terms of the spherical coordinates is converted to the latitude, longitude, and altitude. On the basis of a comparison of this with previously calculated data, (block 362), it is readily possible to determine speed, rate of climb or descent, and true course (block 363) based on successive position calculations.

2.3.5.4 Subscriber Process

The *subscriber process* handles all direct interface with the subscribers, normally (but not necessarily) those who access RDSS data from fixed site (headquarters) locations. The basic process is outlined in the data flow diagram of Figure 2.30. This process establishes the logical links with the subscribers and validates subscriber IDs, based on subscriber profile information passed from the administrative process. Once the subscriber is validated, the process will accept the subscriber's requests for data and incoming keyboard messages, forward them to the store and retrieve process, wait for the appropriate requested data from the store and retreive process, and then forward it to the subscriber.

RDSS greatly lends itself to customized subscriber service. One such service involves specialized reports keyed to position or proximity information. An example of such a report may be a query regarding RDSS system users of a specific class within a defined radius from some point. The subscriber process also records all such transactions and sends these data to the administrative process as event-log entries for billing and accounting. Diagnostic and performance data on the subscriber process itself is also communicated to the administrative process

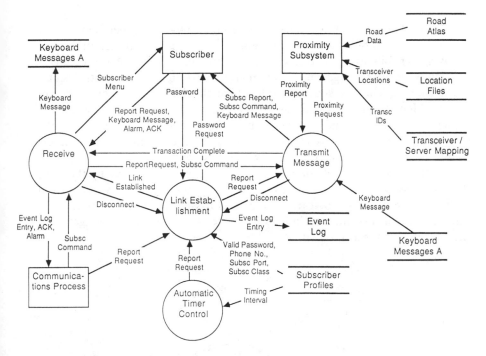

Figure 2.30 Subscriber Server
Source: Geostar Corporation, "International Gateway Station," Contribution
to ITU CITEL Seminar on MOB-WARC-87.

2.4 USER-SEGMENT DESIGN

A standard transceiver design is described in practical detail in this
section. The standard design incorporates a compatibility specification of four
component transceiver "building blocks": the baseband processor, transmitter,
receiver, and serial-device interface. A wide variety of serial devices, such as
keyboard display units (K/D units), may communicate with a RDSS system
through the serial-device interface.

The standard transceiver is the part of a RDSS system that provides
information for precise position determination. The transceiver receives time
reference signals from the control center via a single satellite relay and returns
to the control center signals related to the received time-reference via two or
more satellite relays.

Transceivers may be classified into several categories, depending on user characteristic (mobility and type of application) and the transmitting and receiving capabilities of the transceiver. Refer to Figure 2.31. All transceivers constantly receive the high-rate outbound transmission from the control center via satellite A. This transmission is at a high chip rate (8.000 megachips per second (Mcps)) and contains a ranging code that eventually allows the control center to measure round-trip range by using the transmissions from the transceiver. All users receive the outboard transmissions and look for packets of information addressed to their individual transceivers.

Periodically, at a rate determined by the control center or by the user, the transceivers will burst a response to the time reference mark received from the control center. The burst transmission allows measurement of round-trip range at the control center. Included with the response burst are messages from the user awaiting transmission.

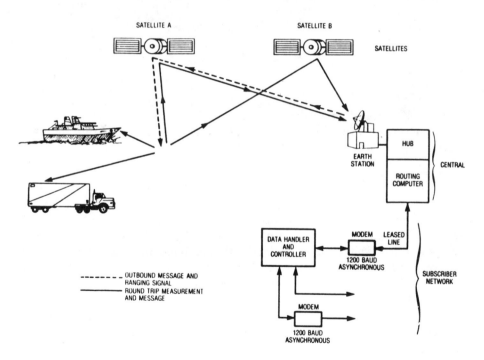

Figure 2.31 RDSS System Overview

A block diagram for the standard transceiver configuration is provided in Figure 2.32. As indicated, a complete transceiver principally consists of the baseband processor, the receiver (Rx chain), and the transmitter (Tx chain).

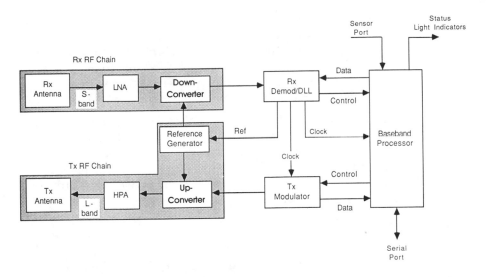

Figure 2.32 Standard Transceiver Configuration

Closely associated with these components are the modulator, demodulator, antenna, and peripheral interfaces. Below we summarize the specifications of the standard transceiver subunits. The summaries are followed by a more detailed explication of the standard information packets that pass among these components.

Baseband Processor (BBP) Overview

The *baseband processor* (BBP) accepts data and status signals from the receiver demodulator and supplies control signals to the receiver and transmitter subunit. The BBP also supplies and receives data to and from the serial port. The major functions of the BBP may be categorized as follows:

Inputs. The BBP receives decoded data from the receiver subunit and user input data for the K/D unit attached to the serial port.

Processing. The BBP manages data flow, acquisition, and all input-output (I/O) functions. The BBP handles all protocol processing for inbound and outbound data, except for the real-time functions of identifying unique words and IDs of the received data and placing preambles in the data for transmission.

Monitor and Control. The BBP maintains the *inhibit timer* and responds to the next PN group after the inhibit timer has gone to zero, thus initiating a return signal, which is used by the control center for position determination.

The BBP also monitors the status of the received signal and enables the transmitting subunit, based on the protocol and the quality of the received signal. The signal quality is monitored on the basis of a field-programmable threshold that enables the transceiver to respond to polls from the control center. Hence, the transceiver will transmit only when it "knows" it has a clear link. This can be reflected on a user display in many ways (e.g., "OK to Send").

Outputs. The BBP provides data outputs via an asynchronous serial communication interface to an external display unit interface. The BBP formats data from the user's serial port input, or from internal sources, to the transmitting subunit. The BBP provides all of the intelligence necessary to perform the functions that enable the transceiver to pass messages back and forth from the serial port to the control center.

Transmitter Overview

The transmitting subunit accepts data from the BBP and timing signals from the receiver chain. By using these signals, the transmitting subunit modulator produces a BPSK signal from a burst input signal of convolutionally encoded data, spread by the 8.000 Mcps of PN sequence. The transmitting subunit then up-converts the signal for transmission and provides power amplification to achieve the specified output power.

Receiver and Delay-Locked Loop (DLL) Overview

The receiver processes a received 2492 MHz RF carrier, which is BPSK modulated, to regenerate the 8.000 Mcps PN signal. The principal functions of the delay-locked loop are acquisition, tracking, synchronization, and decoding. Acquisition provides the receiver chain with initial correlation signals for control of the tracking and synchronization functions. The tracking of the PN sequence provides a clock and an epoch signal to the modulator in order to allow the system to perform coherent loop-back PN ranging. Synchronization enables the recovered BPSK signal to be demodulated by reference to a local clock that is synchronized with the carrier. Decoding provides the *forward error correction* (FEC) function to be performed on the received BPSK data.

2.4.1 Standardized Baseband Processor (BBP) Functions

The transceiver BBP supervises all data I/O transactions, formatting of transmitted messages, deformatting and distribution of received messages, and microprocessor instructings. Interfaces to the BBP include the RF connection to the control center, and asynchronous serial port, and, if desired, a sensor port.

2.4.1.1 Standard Inbound Packet Format

Microprocessor control is essential for transceiver operations, and the microprocessor will ordinarily command immediate execution of the appropriate action when any of several conditions exit. These conditions include:

- Receipt by the BBP of an inbound message of the K/D unit through the serial interface;
- Receipt by the BBP of an indication that there has been a change in the state of the sensor port;
- Removal of primary power or a battery-low condition.

The appropriate action involves assembly of a *standard inbound packet* with prescribed values in a specific RDSS format, such as that depicted in Figure 2.33. The *length byte* defines the message length in a binary representation of the number of bytes in the standard inbound packet, excluding the length, CRC bytes, and the FEC flush bits. The *router address* code distinguishes among different applications and industries that use the RDSS system. This facilitates routing of packets to appropriate computers for processing at one or more system control centers. The 24 bits of this address would ordinarily be stored in a nonvolatile register from which they may be retrieved and modified with ease. The next 48 bits are the *physical address* of the unit, and must be made absolutely unmodifiable. A very large number of physical addresses may be accommodated, and such addresses may be broken down into manufacturer ID, model ID, lot number, and serial number. *Group ID* provides a means for separately processing distinct groups of transceivers by operational or performance characteristics, and the 4 *Ns/Na bits* identify the sequential number of "new" packets, as opposed to repetitions of a packet.

The *application packet* is the message packet that contains information from hardware status, sensor port status, data originating in the serial port such as the K/D unit, or the transceiver itself, as in the case of an automatically timed positioning transmission. The application packet begins with an eight-bit *header*, which is itself comprised of three fields, as shown in Figure 2.34: *priority* (2 bits), *originator domain designator* (ODD, 2 bits), and *originator selector designator* (OSD, 4 bits). Four levels of *priority* are normally accommodated: low, normal, high, and emergency. The ODD describes which of the four areas in an RDSS system, where messages are generated from and

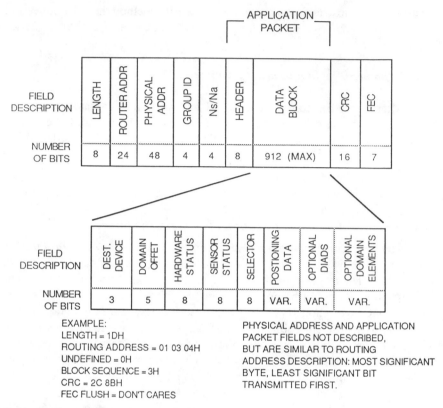

Figure 2.33 Standard Inbound Packet Format
Source: Federal Communications Commission, File No. 1629-DSE-P/L-86 (1986), RDSS Technical Coordinating Committee.

to, has given rise to a particular transmission. The areas are transceiver (00), control center (01), subscriber (10), and administrative (11). The OSS describes which process within a particular domain gave rise to the transmission. The OSS values are unique within each domain. The standardized values for the transceiver domain, in hexadecimal representation, are BBP automatic transmission (1H), receiver acknowledgement of interrogation (2H), encryption module (3H), and K/D unit (4H). This completes the header and brings us to the *data block*.

   The data block is comprised of a number *domain elements*, as shown in Figure 2.34, summing to less than 912 bits in total. The same figure shows that domain elements are themselves composed of *destination designators*, a *domain offset* field, and a number of *diads*, the elemental information unit in an RDSS network. The destination designators (or *device*) is a "send to" address, much like the originator designator functioned as a "return" address. An additional bit is provided for flexibility. The standardized assignments are

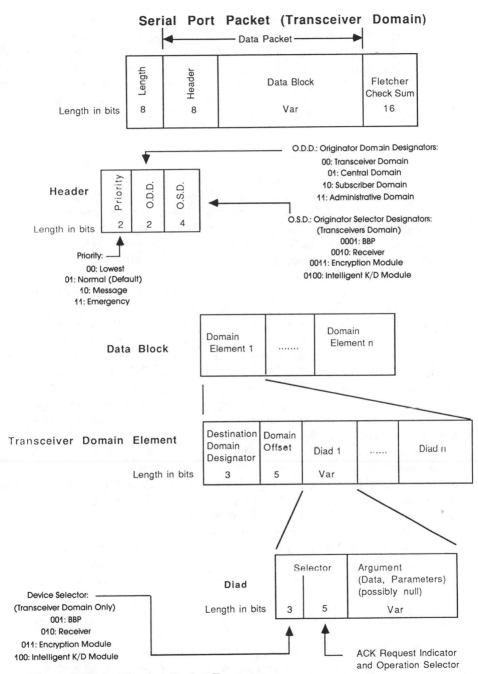

Figure 2.34 Application Packet Format
Source: Federal Communications Commission, File No. 1629-DSE-P/L-86 (1986), RDSS Technical Coordinating Committee.

transceiver (000), control center (001), subscriber (010), administrative (011), and internal to the transmitting transceiver (111). The domain offset is simply a binary representation of the number of bytes in the domain element. This field acts as a pointer to the end of a particular domain element. When set to "zero", the BBP knows it is at the final domain element in a particular data block. A diad is a logical unit comprised of a *selector* and its *argument* (possibly null). Upon receiving a diad, the selector is examined to determine appropriate processing action. The selectors and their arguments are unique for each domain.

Hardware status and *sensor status* bytes, as appropriate, are inserted prior to any diad *by the BBP* (which is why they do not appear in Figure 2.34, a transceiver domain format). The hardware status byte reflects the condition of the transceiver and its attached peripherals. Particular conditions reported are change in sensor port data (a "one" in the least significant bit), failure of transceiver self-check (a "one" in bit place 3), presence of serial ports, and low-battery conditions. The sensor status byte is a representation of its eight-bit contact closure switch positions.

A *selector* byte follows the sensor status byte to represent a three-bit device choice, a one-bit acknowledgement choice, and a four-bit operation choice. The *device selector* targets a subcomponent of the RDSS system to receive and act upon the remainder of the selector and its argument. For the transceiver domain, the standardized device selectors are BBP (001), receiver (010), encryption module (011), and K/D unit (100). The *acknowledgement* (ACK) request bit is exclusively used for the sending application to request that the command to be issued shall be acknowledged by the recipient. This bit is set to "zero" if the sender does not want the command acknowledged, and is set to "one" if the sender desires *transport layer ACK*. The *operation selector* field contains the actual command being issued to the recipient. Commands from the K/D to the BBP are listed in Figure 2.35(a), and an example is provided (Figure 2.35(b)). The argument of a selector is the data, if any, upon which the specified operation is to be performed, and the format and contents of an argument are determined by its associated selector. For automatically generated responses to interrogations, such as timed positioning transmissions, a *positioning data* field follows the selector. This application is illustrated in Figure 2.36. Other frame boundaries and packet definitions are, of course, possible, so long as compatibility and interoperability.

2.4.1.2 Standard Outbound Packet

The BBP must also disassemble outbound packets addressed to it. *Standard outbound packets* are formatted in accordance with high-level data link control (HDLC) standards, as shown in Figure 2.37. The information will

BASEBAND PROCESSOR SELECTOR VALUES

Locally executed BBP commands use the following Selector values:

INITIATOR
0h - NOP
1h - RESET
2h - EXECUTE
3h - AWAKE/ENABLE
4h - ASLEEP/DISABLE
5h - DIAGNOSITCS
6h - READ (Register #)
7h - SET (Register #: value)
8h - BUFFER EMPTY
9h - BUFFER FULL
Ah - CLEAR BUFFER
Bh - Loopback test
Ch - REACQUIRE
Dh -
Eh - Transmit Remainder of
     Data Packet
Fh - Argument contains
     additional Selector codes

RESPONDER
0h - NOP
1h - Reset/Self Test complete
2h - "Executed"
3h - "I am awake"
4h - "I will go to sleep"
5h - "Diagnostics results are:"
6h - "Register # is:"
7h - "Register # set to:"
8h -
9h -
Ah - BUFFER CLEARED
Bh - "In loopback test"
Ch - "Reacquired"
Dh - NAK
Eh -

Fh -

(a)

Initiate a RESET to the BBP

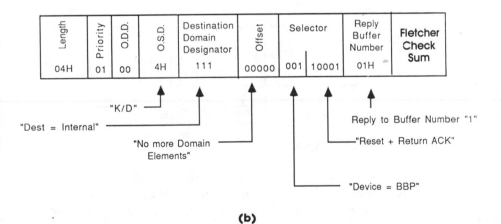

(b)

Figure 2.35 Operation Selector Format:
(a) Commands
(b) Example

PRIORITY	O.D.D.	O.S.D.	D.D.D.	DOMAIN OFFSET	HARDWARE STATUS	SENSOR STATUS	SELECTOR	POSITIONING DATA
01	00	0001	001	00000	XXH	YYH	01H	D'DMMHHD'DDMMHH

O.D.D.: Originator Domain Designator
O.S.D.: Originator Selector Designator
D.D.D.: Destination Domain Designator

Figure 2.36 Automatic Response Application Packet

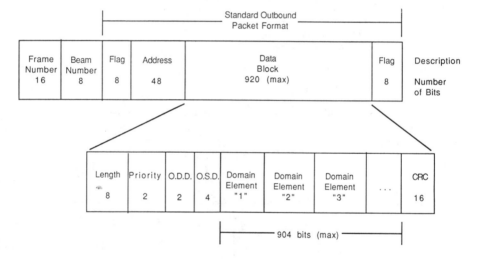

Figure 2.37 Outbound Frame Format

be received in continuously framed formats nominally of 16 ms duration. The 16 data bits following the synchronizing sequence identify the frame with a number that is a running count of the received frames. The frame number must be demultiplexed from the data stream so that the latest number is always stored and updated in the case of inbound transmissions.

The frame number is followed by a beam number, indicating which satellite antenna beam has transmitted a received signal. The beam number is then followed by a message of maximum length, starting with its address. Messages are separated by flags, unique words at the start and end. Several different types of addresses are possible. An immutable physical ID, embedded in the permanent memory (PROM), is essential and should be stored in a register. A match between this address and that in the outbound frame will lead to capture of the message. Other addresses are mnemonic IDs, which can be changed by command from the control center, and group IDs, also alterable by command.

Among the more important of the various BBP registers are the ID register and the *inhibit register*. The inhibit register contains integer values that govern the periodicity of response to interrogation signals. Normal inhibit register values are listed in Table 2.9.

Table 2.9
Inhibit Register Values

Bit	*Units*
7	Inoperable; may be imposed by fail-safe circuit to ensure cessation of transmission.
6	1 day
5	4 hours
4	1 hour
3	10 minutes
2	1 minute
1	30 seconds
0	1 second

Closely related to these inhibit times is the random number generator controlled by the BBP. The random number generator's purpose is to provide a random offset to all timing registers to prevent cascading system-overload phenomena.

2.4.2 Standardized Transmitter Functions

The 1618 MHz transmitter is the pacing element in RDSS transceiver miniaturization efforts. RDSS has increased by many orders of magnitude the production of high-power amplifiers (HPAs) for use at this frequency band. Non-RDSS requirements are less than one thousand per year. The cost of this function will set (within not more than 100%) the manufactured cost of RDSS transceivers.

The transmission to the control center must contain a synchronization code, a PN sequence, and some data. It is important for the data to be convolutionally encoded for error detection and correction and for the PN spreading to occur before BPSK modulation. Once the data are formatted by the BBP into a standard inbound packet, the contents are forwarded at a nominal 16.00 kb/s rate to a recommended constraint-length 7, rate-1/2, convolutional encoder. A block diagram of this encoder is shown in Figure 2.38. The synchronization pattern for the phase-coded acquisition sequence may be similar to that provided in Figure 2.39. Various types of inversions may be used to provide added coding advantage.

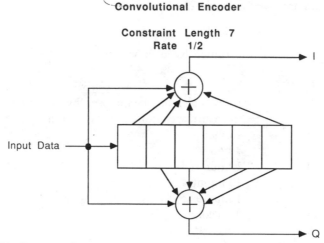

Figure 2.38 Convolutional Encoder

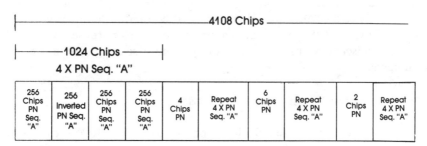

Figure 2.39 Transmitter Signal Synchronization
Source: Federal Communications Commission, RDSS Technical Coordinating Committee (1987).

With a 16 MHz bandwidth, the PN sequence chip rate is nominally 8.000 Mcps. Various patterns may be used to assist the control center in the process of locking onto a code. Several systematic advantages accrue from Gold codes implemented by appropriate pairs of maximal linear codes, where bounded cross-correlation is a key determinant. The selected polynomial pair are hardwired to commands, and the chip rate is controlled by the received outbound clock. With successful synchronization, a gating or "warm-up" signal is sent to the transmitter, and a few milliseconds later the burst transmission may occur. As diagrammed in Figure 2.40, the nominal carrier frequency will be a multiple of the chip rate, subject to any practical variance.

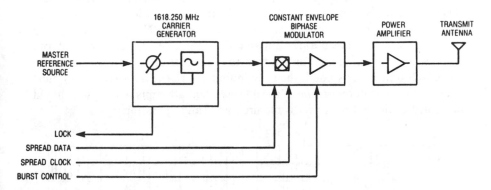

Figure 2.40 RDSS Transmission Chain

2.4.2.1 RF Considerations

Most transmitter RF characteristics may be user-definable due to the flexibility of RDSS Technical Coordinating Committee's *Standard No. 2*. This standard imposes a time-varying statistical power-flux density (PFD) for *all* the transceivers in an RDSS network *at the geostationary orbit*. (See Section 3.2). However, the laws of economics may tend to push many RDSS system operators and users in the direction of lower powered HPAs. In this regard, operators and users must face the realities of satellite communication and space launching capability. The facts of life are that G/T is at high premium, with a figure of 1 dB being the best that we can reasonably expect until close to the year 2000. Finally, boxed in by a gain ceiling of 3 dB for desirably omni-directional antennas and the long distance to the geostationary orbit, we find that standardization activity is occurring in the RF area.

Transmitter EIRP will hover in the 16–22 dBW range at elevation angles of not less than 20° nor greater than 60° or so above the horizon. Transmission must be in the burst mode, according to FCC regulations, with 10ms to 100 ms defining a range of probable reasonableness. Some form of PSK is essential, and BPSK clearly is the standard. Spurious and out-of-band emissions must be carefully dealt with by some sort of filter, due to the $\sin^2 x / x^2$ nature of spread-spectrum power. The FCC regulatory requirements for out-of-band suppression are provided in Section 3.3.3. Careful attention must also be paid to phase noise and jitter in the transmitted signal. See Figure 2.41 for phase noise values that signals are not to exceed. Note that each nanosecond of jitter can subtract a foot from positioning accuracy, with an even worse effect on precision, and it is common to have several nanoseconds of jitter.

Fail-safe protection for the transmitter is an important feature in any RDSS system. Means must exist within the transceiver for transmissions not to occur unless both synchronization to the outbound PN sequence exists and the BBP has commanded a transmission. If any transmission occurs without these conditions being met, or if a transmission extends beyond a maximum value well under one second, then the inhibit register (discussed above) must take on the value of *inoperable* (7). All subsequent attempts to transmit should be denied until the transceiver is repaired and fully operational.

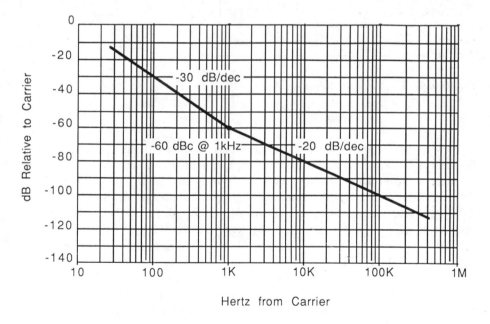

Figure 2.41 Phase Noise of Transmitted RF Signal (Assumed at the Receiver)

2.4.3 Standardized Receiver Specifications

Standard RDSS outbound signal format includes a *preamble* with a synchronization pattern, *spreading* with a PN sequence and BPSK modulation, use of the PN *synchronization clock rate* for transmitting, *convolutional encoding* for error detection and correction, and an *overarching format* structure. These factors, in combination with direct RF concerns, such as receiver compliance with a link budget largely dictated by maximum legal power flux density limits (see Section 2.2), result in standardized receiver specifications.

The receiver's first task is to acquire the outbound synchronization sequence. RDSS systems will normally employ short-cycled (as compared with the standard 2^{17} PN sequence) partial sequence of 256 chips as an acquisition aid. Various techniques are available for maximizing the probability of rapid and accurate acquisition. See Figure 2.42 for an illustrative example of receiver signal-synchronization concept. As a standard, the receiver should acquire in 0.5s or less (99% of the time), when the power-flux density for 2492 MHz is at its nominal design level. Recall that the internationally accepted standard is -144 dBW/m^2, whereas -139 dBW/m^2 is permitted in the United States. A gradual increase in acquisition time is expected over a range of several dB below the minimum specified signal level before there is a complete failure of the acquisition circuit.

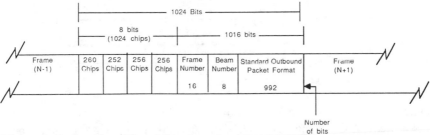

Figure 2.42 Receiver Signal Synchronization
Source: Federal Communications Commission, RDSS Technical Coordinating Committee (1987).

Acquisition has not been accomplished until all of the following have occurred:

- The PN generator is phase locked;
- The carrier recovery loop is phase locked;
- The data bit timing is synchronized.

When the above criteria have been met, the demodulated data should be at a sufficient level for signal detection. As was shown in Figure 2.32, the detected signal is then used to inhibit the acquisition circuit and activate indicators. The clock recovered from the outbound PN sequence is the used to drive the transmitted PN sequence.

Figure 2.43 depicts an exemplary PN sequence generator, based on 17-stage maximal length codes. The despreading of such codes is accomplished by multiplying the received signal by an identical code sequence that is synchronized in time with it. The delay-locked Loop (DLL), referenced in Figure 2.32, balances the outputs of the two despreader circuits operating at IF. The despreaders are each driven by two PN code inputs, one of which is one-half of a chip early and the other is one-half of a chip late. The lock-up process is shown in Figure 2.44.

PN GENERATOR $(X^{17} + X^{14} + X^{11} + X^{10} + X^8 + X^7 + X^3 + X + 1)$

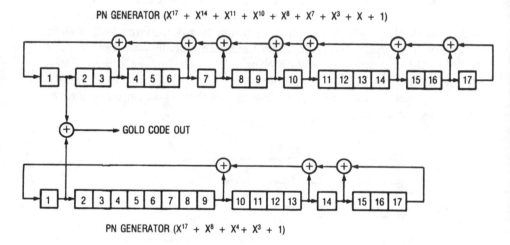

PN GENERATOR $(X^{17} + X^8 + X^4 + X^3 + 1)$

Figure 2.43 Gold Code PN Generator
Source: Federal Communications Commission Assignment to Geostar Corporation, File No. 1629-DSE-P/L-86 (1987).

In order to achieve the standardized BER performance of 10^{-5} within the nominal RDSS link budgets, it is also necessary to employ a decoder that is capable of decoding a constraint-length 7, rate-1/2, convolutionally encoded signal. The coding gain that results from such FEC encoding is shown in Figure 2.45. This decoding capability is commercially available on very-large-scale integration (VLSI) Viterbi decoding chips. As explained in Section 2.1.2, coding gain is essential for meeting RDSS link budget requirements.

2.4.4 Standardized Serial Device Interface

We noted earlier that RDSS transceivers will ordinarily include, at minimum, serial and sensor device ports. The serial device port is especially important because it is through this interface that keyboard-display units can

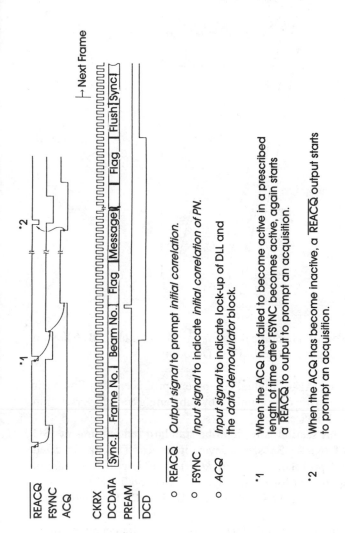

○ REACQ   *Output signal* to prompt *initial correlation.*

○ FSYNC   *Input signal* to indicate *initial correlation of PN.*

○ ACQ   *Input signal* to indicate lock-up of DLL and
           the *data demodulator* block.

*1   When the ACQ has failed to become active in a prescribed
           length of time after FSYNC becomes active, again starts
           a REACQ to output to prompt an acquisition.

*2   When the ACQ has become inactive, a REACQ output starts
           to prompt an acquisition.

○ PREAM   *Input signal* to latch *frame no.* and *beam no.,*
           at the same time, notifies CPU of the
           latch concluded.

Figure 2.44 Delay-Locked Loop Receiver Process

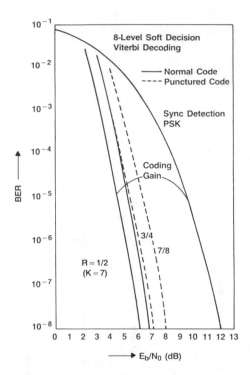

Figure 2.45 Comparison of Theoretical E_b/N_0 for Demodulated BPSK with Various Rates of Soft-Decision Viterbi FEC Decoding

access RDSS capabilities. Also, original-equipment producers will find their greatest RDSS manufacturing opportunities by subsuming transceivers within market-specialized serial devices.

The purpose of this section is to describe standardized features at ISO Layer 2 (*link layer*) of the RDSS serial device interface. The standardized features of ISO Level 3 protocol (*transport layer*) were described as part of the standard serial-port packet format discussion in Section 2.4.1. Electromechanical interface is RS-232 for the serial port and SAE J1455 for applications in heavy duty vehicles.

The serial communication interface includes two independent data-transfer channels, one from the serial port devices to the baseband processor and the other from the baseband processor to the serial port devices. Standardization of link-level protocol for these channels enables an open market in RDSS electronics. The channels use identical protocols to accomplish data transfer. Each channel will consist of a transmitter (XMTR) and a receiver (RCVR) connected by three circuits: a DATA circuit, a READY circuit, and a PRESENT circuit. The RCVR controls the READY circuit, and the XMTR controls the PRESENT and DATA circuits. Each PRESENT circuit indicates that its source (either the serial port device or the BBP) is plugged in, powered on, and operational (subject to diagnostics provided in the equipment). Any data arriving from a XMTR would be disregarded if its associated PRESENT is absent. The READY circuit indicates to the XMTR that the RCVR has the necessary resources (processor capacity and buffer space) to accept a maximum-length message.

The first byte sent by the XMTR to the RCVR following assertion of READY must be the length byte that starts a message. Successive message bytes are then transferred, ending with Fletcher check-sum bytes. READY must be asserted throughout the transfer and should be checked by the XMTR, either continuously or prior to transmission of each character. Then, following receipt of the length byte plus the number of characters indicated and two check-sum bytes, the RCVR must drop READY for at least 1.5 byte periods. This provides an opportunity for XMTR to note that READY has dropped by the RCVR.

After the check sum, the XMTR continues to check READY and sends characters consisting of binary "zeros" until READY is dropped by the RCVR. This action ensures that the RCVR will reach an end-of-message condition, even if the length byte was damaged by an error and appeared to be too large. Transmission of "zeros" will also allow recoordination of the XMTR and the RCVR if the XMTR has failed to note a drop and reassertion of READY because it will look like the beginning of a zero-length message, causing the RCVR to drop READY again. Of course, the XMTR must cease transmission upon noting any drop of READY.

The foregoing protocol provides message definition (where does a message begin and end), pacing (is the receiver capable of receiving), and transmission error checking (Fletcher check-sum approach). The higher level transport-layer protocol used between the serial port device and the BBP to ensure an orderly exchange and execution of commands and responses was depicted in Section 2.4.1. A link-level ACK/NAK is the response from the RCVR to the XMTR for the status of a message block reception. The length of a link-level ACK/NAK is three bytes: the ACK/NAK byte and two Fletcher check-sum bytes. A link-level ACK/NAK is sent to the XMTR, depending on the validity of the length and check-sum bytes. If either is found to be incorrect, a link-level NAK (code=80H) is returned to the transmitter. An ACK (code = F0H) is returned otherwise. Absence of buffer space would also result in a NAK.

2.4.5 Nontransceiver-Based Interface to Control Center

Often, transceiver transmissions to the control center are retransmitted by the control center over non-RDSS transmission paths and messages to transceivers are received by the control center over non-RDSS links. For example, a large transport company ("carrier") would ordinarily wish that periodic RDSS position reports on its vehicles would be sent to the company's dispatch center, the headquarters of the entity that contracted for the transport of its good ("shipper"), the intended destination of the goods, the company's insurers, and various regulatory bodies. Although such multiply addressed transmissions could be sent over the outbound RDSS circuit, it will often be more cost-effective to route them via the switched telephone network or the wide variety of private data and packet-switched networks. The reader should note that such techniques also directly increase RDSS capacity because it is the outbound channel which limits capacity in all cases.

There is little need for RDSS-specific standardization in the RDSS to non-RDSS transmission interface. Each RDSS system will incorporate protocols that are optimal for the users of its system. Commercially available RDSS *user headquarters software* is available for the IBM PC/AT and MacIntosh family of computers. This software is designed to receive positioning reports (latitude and longitude) and messages from transceivers via an RDSS control center. The software then stores position, message, transceiver ID number, and time in a data-base format from which it may be graphically displayed as an overlay on a digitally stored map of the US interstate highway system. We should note that the US interstate highway system, by far the world's most extensive, may be digitally stored in less than 2 Mbytes.

Appendix 2.A
RDSS Accuracy Matrix
Calculations (3 × 7)

The following pages contain the three-by-seven (3×7) sensitivity matrices $S(n)$ expressed in earth centered inertial (ECI) coordinates and local east, north, and up coordinates for each of 12 nominal user locations.

3 × 7 SENSITIVITY MATRICES

SATELLITE A AT LONGITUDE   −74.000
SATELLITE B AT LONGITUDE −131.000

USER LONGITUDE = −90.000
USER LATITUDE =   25.000
USER ALTITUDE ABOVE REFERENCE SPHERE =   .000

THE X,Y,Z SENSITIVITY MATRIX, BY COLUMNS TRANSPOSED

−.77274E+00	.67173E+00	.14405E+01	$\Delta\rho_A$
.10250E+01	.29392E+00	.63032E+00	$\Delta\rho_B$
.76922E+00	−.66868E+00	−.14340E+01	ΔR_A
−.10155E+01	−.29119E+00	−.62446E+00	ΔR_B
.17671E+04	−.15361E+04	−.32943E+04	$\Delta\theta_A$
−.33445E+04	−.95901E+03	−.20566E+04	$\Delta\theta_B$
−.41048E−01	−.18515E+00	.19691E+01	Δh

THE E,N, UP SENSITIVITY MATRIX, BY COLUMNS TRANSPOSED

−.77274E+00	.15895E+01	.11102E−15	$\Delta\rho_A$
.10250E+01	.69548E+00	.00000E+00	$\Delta\rho_B$
.76922E+00	−.15822E+01	−.11102E−15	ΔR_A
−.10155E+01	−.68902E+00	.00000E+00	ΔR_B
.17671E+04	−.36348E+04	.00000E+00	$\Delta\theta_A$
−.33445E+04	−.22692E+04	.11369E−12	$\Delta\theta_B$
−.41048E−01	.17064E+01	.10000E+01	Δh

SATELLITE A AT LONGITUDE   -74.000
SATELLITE B AT LONGITUDE   -131.000

USER LONGITUDE   -90.000
USER LATITUDE   30.000
USER ALTITUDE ABOVE REFERENCE SPHERE   .000

THE X,Y,Z SENSITIVITY MATRIX, BY COLUMNS TRANSPOSED

-.77947E+00	.67759E+00	.11736E+01
.10315E+01	.29578E+00	.51230E+00
.77503E+00	-.67373E+00	-.11669E+01
-.10213E+01	-.29286E+00	-.50724E+00
.19944E+04	-.17337E+04	-.30028E+04
-.34699E+04	-.99497E+03	-.17233E+04
-.41048E-01	-.18515E+00	.16793E+01

THE E,N,UP SENSITIVITY MATRIX, BY COLUMNS TRANSPOSED

-.77947E+00	.13552E+01	-.11102E-15
.10315E+01	.59156E+00	-.11102E-15
.77503E+00	-.13475E+01	.00000E+00
-.10213E+01	-.58572E+00	.55511E-16
.19944E+04	-.34674E+04	.22737E-12
-.34699E+04	-.19899E+04	.11369E-12
-.41048E-01	.13617E+01	.10000E+01

SATELLITE A AT LONGITUDE   -74.000
SATELLITE B AT LONGITUDE   -131.000

USER LONGITUDE   -90.000
USER LATITUDE   45.000
USER ALTITUDE ABOVE REFERENCE SPHERE   .000

THE X,Y,Z SENSITIVITY MATRIX, BY COLUMNS TRANSPOSED

-.80549E+00	.70020E+00	.70020E+00
.10567E+01	.30299E+00	.30299E+00
.79794E+00	-.69364E+00	-.69364E+00
-.10442E+01	-.29943E+00	-.29943E+00
.26402E+04	-.22951E+04	-.22951E+04
-.38773E+04	-.11118E+04	-.11118E+04
-.41048E-01	-.18515E+00	.12291E+01

THE E,N,UP SENSITIVITY MATRIX, BY COLUMNS TRANSPOSED

-.80549E+00	.99024E+00	-.11102E-15
.10567E+01	.42849E+00	-.55511E-16
.79794E+00	-.98096E+00	.16653E-15
-.10442E+01	-.42345E+00	.27756E-16
.26402E+04	-.32457E+04	.45475E-12
-.38773E+04	-.15723E+04	.11369E-12
-.40148E-01	.73815E+00	.10000E+01

SATELLITE A AT LONGITUDE   −74.000
SATELLITE B AT LONGITUDE   −131.000

USER LONGITUDE   −90.000
USER LATITUDE   75.000
USER ALTITUDE ABOVE REFERENCE SPHERE   .000

THE X,Y,Z SENSITIVITY MATRIX, BY COLUMNS TRANSPOSED

−.87472E+00	.76038E+00	.20374E+00
.11246E+01	.32247E+00	.86405E−01
.86257E+00	−.74982E+00	−.20091E+00
−.11089E+01	−.31796E+00	−.85197E−01
.34864E+04	−.30307E+04	−.81206E+03
−.44964E+04	−.12893E+04	−.34547E+03
−.41048E−01	−.18515E+00	.98566E+00

THE E,N,UP SENSITIVITY MATRIX, BY COLUMNS TRANSPOSED

−.87472E+00	.787221E+00	.00000E+00
.11246E+01	.33384E+00	.00000E+00
.86257E+00	−.77627E+00	−.27756E−16
−.11089E+01	−.31918E+00	−.13878E−16
.34864E+04	−.31376E+04	−.11369E−12
−.44964E+04	−.13348E+04	.00000E+00
−.41048E−01	.76264E−01	.10000E+01

SATELLITE A AT LONGITUDE   −74.000
SATELLITE B AT LONGITUDE   −131.000

USER LONGITUDE   −78.000
USER LATITUDE   40.000
USER ALTITUDE ABOVE REFERENCE SPHERE   .000

THE X,Y,Z SENSITIVITY MATRIX, BY COLUMNS TRANSPOSED

−.79121E+00	.687679E+00	.99781E+00
.10716E+01	.30728E+00	.92675E−01
.78528E+00	−.68263E+00	−.99032E+00
−.10581E+01	−.30341E+00	−.91508E−01
.23217E+04	−.20182E+04	−.29280E+04
−.40684E+04	−.11666E+04	−.35185E+03
.41048E−01	−.18515E+00	.13501E+01

THE E,N,UP SENSITIVITY MATRIX, BY COLUMNS TRANSPOSED

−.63092E+00	.13025E+01	.11102E−15
.11121E+01	.12098E+00	.20817E−16
.62619E+00	−.12928E+01	.00000E+00
−.10981E+01	−.11946E+00	−.20817E−16
.18514E+04	−.38222E+04	−.22737E−12
−.42221E+04	−.45930E+03	−.11369E−12
−.78646E−01	.92328E+00	.10000E+01

SATELLITE A AT LONGITUDE   -74.000
SATELLITE B AT LONGITUDE -131.000

USER LONGITUDE -85.000
USER LATITUDE   40.000
USER ALTITUDE ABOVE REFERENCE SPHERE   .000

THE X,Y,Z SENSITIVITY MATRIX, BY COLUMNS TRANSPOSED

-.79329E+00	.68960E+00	.90110E+00
.10570E+01	.30308E+00	.25004E+00
.78711E+00	-.68422E+00	-.89408E+00
-.10445E+01	-.29951E+00	-.24709E+00
.23728E+04	-.20627E+04	-.26953E+04
-.38817E+04	-.11130E+04	-.91825E+03
-.41048E-01	-.18515E+00	.13402E+01

THE E,N,UP SENSITIVITY MATRIX, BY COLUMNS TRANSPOSED

-.73017E+00	.11763E-01	-.11102E-15
.10794E+01	.32640E+00	.00000E+00
.72448E+00	-.11671E+01	.11102E-15
-.10666E+01	-.32256E+00	.00000E+00
.21840E+04	-.35184E+04	.45475E-12
-.39639E+04	-.11987E+04	.00000E+00
-.57029E-01	.91037E+00	.10000E+01

SATELLITE A AT LONGITUDE   -74.000
SATELLITE B AT LONGITUDE -131.000

USER LONGITUDE = -100.000
USER LATITUDE =   40.000
USER ALTITUDE ABOVE REFERENCE SPHERE   .000

THE X,Y,Z SENSITIVITY MATRIX, BY COLUMNS TRANSPOSED

-.80401E+00	.69892E+00	.65390E+00
.10309E+01	.29559E+00	.56025E+00
.79662E+00	-.69249E+00	-.64789E+00
-.10207E+01	-.29269E+00	-.55475E+00
.26105E+04	-.22692E+04	-.21231E+04
-.34578E+04	-.99152E+03	-.18793E+04
-.41048E-01	-.18515E+00	.13299E+01

THE E,N,UP SENSITIVITY MATRIX, BY COLUMNS TRANSPOSED

-.91317E+00	.85360E+00	-.16653E-15
.96387E+00	.73136E+00	-.11102E-15
.90477E+00	-.84576E+00	.11102E-15
-.95441E+00	-.72418E+00	.11102E-15
.29649E+04	-.27715E+04	.00000E+00
-.32331E+04	-.24532E+04	.22737E-12
-.82724E-02	.89699E+00	.10000E+01

SATELLITE A AT LONGITUDE   −74.000
SATELLITE B AT LONGITUDE   −131.000

USER LONGITUDE   −125.000
USER LATITUDE   40.000
USER ALTITUDE ABOVE REFERENCE SPHERE   .000

THE X,Y,Z SENSITIVITY MATRIX, BY COLUMNS TRANSPOSED

−.83795E+00	.72842E+00	.13831E+00
.10083E+01	.28911E+00	.97145E+00
.82758E+00	−.71941E+00	−.13660E+00
−.10006E+01	−.28693E+00	−.96411E+00
.31537E+04	−.27414E+04	−.52054E+03
−.29698E+04	−.85157E+03	−.28613E+04
−.41048E−01	−.18515E+00	.13469E+01

THE E,N,UP SENSITIVITY MATRIX, BY COLUMNS TRANSPOSED

−.11042E+01	.18055E+00	−.55511E−16
.66009E+00	.12681E+01	.00000E+00
.10906E+01	−.17832E+00	.41633E−16
−65510E+00	−.12586E+01	.00000E+00
.41557E+04	−.67952E+03	.22737E−12
−.19442E+04	−.37352E+04	.00000E+00
.72575E−01	.91917E+00	.10000E+01

SATELLITE A AT LONGITUDE   −74.000
SATELLITE B AT LONGITUDE   −131.000

USER LONGITUDE   −90.000
USER LATITUDE   30.000
USER ALTITUDE ABOVE REFERENCE SPHERE   .000

THE X,Y,Z SENSITIVITY MATRIX, BY COLUMNS TRANSPOSED

−.80549E+00	.70020E+00	.70020E+00
.10566E+01	.30299E+00	.30299E+00
.79794E+00	−.69364E+00	−.69364E+00
−.10442E+01	−.29943E+00	−.29943E+00
.26403E+04	−.22951E+04	−.22951E+04
−.38774E+04	−.11118E+04	−.11118E+04
−.41049E−01	−.18516E+00	.12291E+01

THE E,N,UP SENSITIVITY MATRIX, BY COLUMNS TRANSPOSED

−.80549E+00	.99023E+00	.00000E+00
.10566E+01	.42849E+00	−.27756E−16
.79794E+00	−.98095E+00	.00000E+00
−.10442E+01	−.42345E+00	.27756E−16
.26403E+04	−.32458E+04	.00000E+00
−.38774E+04	−.15723E+04	.00000E+00
−.41049E−01	.73815E+00	.10000E+01

SATELLITE A AT LONGITUDE   −74.000
SATELLITE B AT LONGITUDE   −131.000

USER LONGITUDE   −90.000
USER LATITUDE   45.000
USER ALTITUDE ABOVE REFERENCE SPHERE   .200
THE X,Y,Z SENSITIVITY MATRIX, BY COLUMNS TRANSPOSED

−.80549E+00	.70020E+00	.70020E+00
.10566E+01	.30299E+00	.30299E+00
.79794E+00	−.69364E+00	−.69364E+00
−.10442E+01	−.29943E+00	−.29943E+00
.26403E+04	−.22952E+04	−.22952E+04
−.38775E+04	−.11118E+04	−.11118E+04
−.41050E−01	−.18516E+00	.12291E+01

THE E,N,UP SENSITIVITY MATRIX, BY COLUMNS TRANSPOSED

−.80549E+00	.99023E+00	−.11102E−15
.10566E+01	.42849E+00	.27756E−16
.79794E+00	−.98095E+00	.16653E−15
−.10442E+01	−.42345E+00	.27756E−16
.26403E+04	−.32459E+04	.22737E−12
−.38775E+04	−.15724E+04	.11369E−12
−.41050E−01	.73814E+00	.10000E+01

SATELLITE A AT LONGITUDE   −74.000
SATELLITE B AT LONGITUDE   −131.000

USER LONGITUDE   −90.000
USER LATITUDE   45.000
USER ALTITUDE ABOVE REFERENCE SPHERE   .300
THE X,Y,Z SENSITIVITY MATRIX, BY COLUMNS TRANSPOSED

−.80548E+00	.70020E+00	.70020E+00
.10566E+01	.30299E+00	.30299E+00
.79794E+00	−.69364E+00	−.69364E+00
−.10442E+01	−.29943E+00	−.29943E+00
.26404E+04	−.22953E+04	−.22953E+04
−.38776E+04	−.11119E+04	−.11119E+04
−.41051E−01	−.18517E+00	.12290E+01

THE E,N,UP SENSITIVITY MATRIX, BY COLUMNS TRANSPOSED

−.80548E+00	.99023E+00	−.55511E−16
.10566E+01	.42849E+00	.00000E+00
.79794E+00	−.98095E+00	.11102E−15
−.10442E+01	−.42345E+00	.00000E+00
.26404E+04	−.32460E+04	.45475E−12
−.38776E+04	−.15724E+04	.11369E−12
−.41051E−01	.73813E+00	.10000E+01

SATELLITE A AT LONGITUDE   –74.000
SATELLITE B AT LONGITUDE   –131.000

USER LONGITUDE   –90.000
USER LATITUDE   45.000
USER ALTITUDE ABOVE REFERENCE SPHERE   .400

THE X,Y,Z SENSITIVITY MATRIX, BY COLUMNS TRANSPOSED

–.80548E+00	.70020E+00	.70020E+00
.10566E+01	.30299E+00	.30299E+00
.79793E+00	–.69363E+00	–.69363E+00
–.10442E+01	–.29942E+00	–.29942E+00
.26405E+04	–.22953E+04	–.22953E+04
–.38777E+04	–.11119E+04	–.11119E+04
–.41052E–01	–.18517E+00	.12290E+01

THE E,N,UP SENSITIVITY MATRIX, BY COLUMNS TRANSPOSED

–.80548E+00	.99023E+00	.55511E–16
.10566E+01	.42849E+00	.27756E–16
.79793E+00	–.98094E+00	–.11102E–15
–.10442E+01	–.42345E+00	.00000E+00
.26405E+04	–.32461E+04	.00000E+00
–.38777E+04	–.15725E+04	.00000E+00
–.41052E–01	.73813E+00	.10000E+01

Appendix 2.B
RDSS Accuracy Matrix
Calculations (3 × 9)

The following pages contain the three-by-nine (3 × 9) sensitivity matrices S (n) for position determination using three satellites. Four different nominal user locations are considered here.

3 × 9 SENSITIVITY MATRICES

SATELLITE A AT LONGITUDE -74.000
SATELLITE B AT LONGITUDE -131.00
SATELLITE C AT LONGITUDE -105.00

USER LONGITUDE = -90.000
USER LATITUDE = 45.000
USER ALTITUDE ABOVE REFERENCE SPHERE = .000
THE X,Y,Z SENSITIVITY MATRIX, BY COLUMNS TRANSPOSED

-16477E+01	-.30990E+01	.25919EE+02
.37475E-01	-.42942E+01	.30820E+02
.16103E+01	.72634E+01	-.48215E+02
.16323E+01	.30699E+01	-.25676E+02
-.37035E-01	.42437E+01	-.30457E+02
-.15953E+01	-.71958E+01	.47766E+02
.54009E+04	.10158E+05	-.84957E+05
-.13751E+03	.15757E+05	-.11309E+06
-.52595E+04	-.23724E+05	.15748E+06

THE E,N,UP SENSITIVITY MATRIX, BY COLUMNS TRANSPOSED

-.16477E+01	.16136E+02	.20519E+02
.37475E-01	.18756E+02	.24829E+02
.16103E+01	-.28957E+02	-.39229E+02
.16323E+01	-.15985E+02	-.20327E+02
-.37034E-01	-.18536E+02	-.24537E+02
-.15953E+01	.28688E+02	.38864E+02
.54009E+04	-.52891E+05	-.67256E+05
-.13751E+03	-.68824E+05	-.91108E+05
-.52595E+04	.94581E+05	.12813E+06

SATELLITE A AT LONGITUDE   -74.000
SATELLITE B AT LONGITUDE -131.000
SATELLITE C AT LONGITUDE -105.000

USER LONGITUDE -90.000
USER LATITUDE   75.000
USER ALTITUDE ABOVE REFERENCE SPHERE   .000

THE X,Y,Z SENSITIVITY MATRIX, BY COLUMNS TRANSPOSED

-.17894E+01	-.33653E+01	.22167E+02
.39884E-01	-.45703E+01	.26133E+02
.17495E+01	.78914E+01	-.42010E+02
.17645E+01	.33186E+01	-.21859E+02
-.39327E-01	.45064E+01	-.25768E+02
-.17252E+01	-.77818E+01	.41427E+02
.71319E+04	.13413E+05	-.88350E+05
-.15947E+03	.18273E1+05	-.10449E+06
-.69721E+04	-.31449E+05	.16742E+06

THE E,N,UP SENSITIVITY MATRIX, BY COLUMNS TRANSPOSED

-.17894E+01	.24866E+01	.22283E+02
.39884E-01	.23491E+01	.26425E+02
.17495E+01	-.32504E+01	-.42621E+02
.17645E+01	-.24520E+01	-.21973E+02
-.39327E-01	-.23163E+01	-.26056E+02
-.17252E+01	.32053E+01	.42029E+02
.71319E+04	-.99107E+04	-.88811E+05
-.15947E+03	-.93926E+04	-.10566E+06
-.69721E+04	.12954E+05	.16986E+06

SATELLITE A AT LONGITUDE   -74.000
SATELLITE B AT LONGITUDE -131.000
SATELLITE C AT LONGITUDE -105.000

USER LONGITUDE −90.000
USER LATITUDE   25.000
USER ALTITUDE ABOVE REFERENCE SPHERE   .000
THE X,Y,Z SENSITIVITY MATRIX, BY COLUMNS TRANSPOSED

−.15807E+01	−.29730E+01	.40202E+02
.36354E−01	−.41657E+01	.48059E+02
.15444E+01	.69662E+01	−.74086E+02
.15736E+01	.29594E+01	−.40020E+02
−.36016E−01	.41270E+01	−.47613E+02
−.15375E+01	−.69354E+01	.73759E+02
.36149E+04	.67987E+04	−.91936E+05
−.11861E+03	.13592E+05	−.15681E+06
−.34805E+04	−.15700E+05	.16697E+06

THE E,N,UP SENSITIVITY MATRIX, BY COLUMNS TRANSPOSED

−.15807E+01	.35179E+02	.19685E+02
.36353E−01	.41796E+02	.24086E+02
.15444E+01	−.64201E+02	−.37624E+02
.15736E+01	−.35019E+02	−.19595E+02
−.36016E−01	−.41408E+02	−.23862E+02
−.15375E+01	.63917E+02	.37458E+02
.36149E+04	−.80449E+05	−.45016E+05
−.11861E+03	−.13637E+06	−.78588E+05
−.34805E+04	.14469E+06	.84792E+05

SATELLITE A AT LONGITUDE   −74.000
SATELLITE B AT LONGITUDE −131.000
SATELLITE C AT LONGITUDE −105.000

USER LONGITUDE −90.000
USER LATITUDE   30.000
USER ALTITUDE ABOVE REFERENCE SPHERE   .000
THE X,Y,Z SENSITIVITY MATRIX, BY COLUMNS TRANSPOSED

−.15945E+01	−.29989E+01	.34518E+02
.36583E−01	−.41920E+01	.41216E+02
.15579E+01	.70273E+01	−.63736E+02
.15854E+01	.29818E+01	−.34322E+02
−.36222E−01	.41506E+01	−.40809E+02
−15492E+01	−.69881E+01	.63380E+02
.40798E+04	.76729E+04	−.88319E+05
−.12306E+03	.14102E+05	−.13865E+06
−.39456E+04	−.17798E+05	.16142E+06

THE E,N,UP SENSITIVITY MATRIX, BY COLUMNS TRANSPOSED

-.15945E+01	.28394E+02	.19856E+02
.36583 E-01	.33598E+02	.24238E+02
.15579E+01	-.51683E+02	-.37954E+02
.15854E+01	-.28232E+02	-.19743E+02
-.36222E-01	.51395E+02	-.23999E+02
-.15492E+01	.51395E+02	.37742E+02
.40798E+04	-.72650E+05	-.50804E+05
-.12306E+03	-.11302E+06	-.81535E+05
-.39456E+04	.13089E+06	.96123E+05

Chapter 3
System Management

3.1 MANAGEMENT OF SYSTEM TRAFFIC

This chapter considers the issues of management of an RDSS system's own traffic, simultaneous operation of more than one RDSS system in the same coverage area, and controlling interference between RDSS and non-RDSS systems. The random-access nature of RDSS — users transmit whenever they please, without channel assignment — implies that management of RDSS system traffic should not be too complex. This is only true, however, if there is adequate system capacity. The random-access system *without* adequate capacity is an *unmanageable* system.

The twin questions of interference within the RDSS service and with non-RDSS radio systems is also very important. These interference considerations had a major effect in shaping the technical parameters of RDSS. Such considerations and their technical or regulatory "fixes" are also discussed below.

3.1.1 Random-Access Spread-Spectrum TDMA Protocol

The use of spread-spectrum modulation in conjunction with time-division multiplexing (TDM) techniques provides several benefits as compared to other, more standard, TDM implementations. These benefits include greater capacity for high-precision ranging and simpler satellite and user transceiver design. Although random access is not a necessary feature of RDSS, the spread-spectrum overlay allows enjoyment of the simplicity and flexibility of random access without the low-capacity ceiling for which such access schemes are infamous.

A spread-spectrum TDM communication system can have a capacity greater than either an *ALOHA*, or even nonspread-spectrum *slotted ALOHA*, TDM scheme. ALOHA channels, first characterized by Dr. Norman Abrahamson of the University of Hawaii, are common frequency channels accessed by users on a random basis. We are able to show mathematically that because of channel saturation, due to the cascading effects of users seeking to retransmit, an ALOHA channel cannot exceed 18% of its maximum Shannon information capacity. [See CCIR *Report 741*.] A slotted ALOHA scheme is one in which random access still prevails, subject to the condition that any access correspond to specific periodic time intervals. We may see that this doubles channel capacity to 36% of the theoretical limit. RDSS incorporates a slotted ALOHA scheme because user transmissions must correspond to specific interrogation epochs, or "frames," as described in Section 2.3 (see Figure 2.20).

The spread-spectrum nature of RDSS enables it to exceed the slotted ALOHA ceiling of 36% by a factor of 1.7. Spread-spectrum slotted ALOHA can achieve, on a normalized basis, 60% of the theoretical channel capacity because message collisions in time no longer necessarily cause lost information. Recall from Section 2.1.3 that spread-spectrum modulation provides a bit substructure of chips, which, if lost, can be reconstructed from knowledge of the chip-generating code. A collision of chips in a spread-spectrum system merely degrades the received energy per bit to noise ratio. Such degradation can occur in the system up to the limit of margin before information throughput declines. Channel degradation is graceful rather than catastrophic. In general, if two spread-spectrum signals collide in the receiver, they may still be correctly received if they were transmitted by using PN codes with good cross-correlation properties. If two signals using the same PN code collide in the receiver, it may still be possible to decode both messages, provided that there are two decoders available for the PN code, or the two signals are slightly offset in time.

Spread-spectrum modulation also allows the implementation of both the user transceiver and satellite-to-user transmitters without the use of extremely high-peak-power devices. This is because coding techniques involve a trade-off of peak power for bandwidth. To implement RDSS without the roughly 128-to-1 coding gain described in the CCIR reference system would necessitate increasing transceiver power and satellite G/T by about 10 dB *each*, or drastically reducing the system's data rate. This pushes RDSS to, if not beyond, the limit of economic practicality through at least the mid-1990s.

3.1.2 System Loading Constraints

A workable estimate of system capacity can be obtained from the CCIR's RDSS reference-system link budgets provided in Section 2.2. Recall that sufficient margin existed to support an information rate, after FEC coding, of 64 kb/s on the outbound channel and 16 kb/s on the inbound channel. We necessarily assume a usage model for the data rates to be turned into estimated user loading constraints on the system.

In general, the outbound link capacity is modeled by

Outbound Capacity = (3600) × (Frame Rate)
× (Average Messages/Frame) per Hour

The CCIR's RDSS reference system specifies a frame period of 12 ms, and hence the frame rate is 83.3 per second. Assume each frame is packed, on the average, with two messages. Then, the outbound capacity is 600,000 transmissions per hour. This figure should be multiplied by an *efficiency factor* to account for framing set-up, repetitions, and queueing. By setting this factor at 0.76 = (0.95 framing efficiency 0.80 repetition/queue efficiency), we obtain an expected capacity of about 450,000 outbound messages per hour.

The inbound link capacity is driven by the number of simultaneous messages that can be received and the average message length. The number of simultaneous messages K may be derived from the following formula:

$$(E_b/N_{0K})^{-1} = (E_b/N_{01})^{-1} + 2/3[(K-1)(T_b/T_c)^{-1}]$$

where

$[E_b/N_{01}]$ = signal to noise for a single signal, about 6 dB including margins;

$[T_b/T_c]$ = spreading gain = 27 dB with bit duration T_b, derived from 16 kb/s; and chip duration T_c derived from 8 Mbps;

$[E_b/N_{0K}]$ = required 4 dB Viterbi threshold for BER of 10^{-5}.

By solving for K, we obtain K = 114 simultaneous messages.

Assuming that each inbound transmission is 24 ms in length (see Figure 2.19), we can derive an inbound loading limit of 17 million accesses per hour. Therefore, it should be clear that the loading constraint in an RDSS system is in the outbound link, assuming that we have sufficient supply of acquisition units and decoders at the control center to handle the link capacity.

3.1.3 Data Processing Requirements

Data processing requirements in an RDSS system include:

a. acquisition and processing of transmissions from mobiles;
b. processing of electronic mail transmissions from fixed sites;
c. preparation and sending of transmissions to mobiles;
d. preparation and transmission of electronic mail to fixed sites;
e. control of the system and the payloads;
f. traffic control;
g. statistics and invoicing;
h. customer services, such as detection of incipient terminal failures.

Figures 3.1 (software) and 3.2 (hardware) diagram the data processing functions required in an RDSS system. After decoding, a routing computer must be able to handle a digital data stream at its maximum expected rate. Such a computer, normally in a distributed local area network architecture, would pass off received packets to separate computers, based on updated routing tables stored in fast access memory. These separate processors would examine the message structure to determine whether it constituted a position request, a message to be relayed, or an acknowledgement. If terrain or other data-base access were required, the packet would be handed off to a specialized routine written for such fast data-base access and computation. After processing, a system access count is registered on a billing or statistics machine and an outbound packet is structured for encoding, if appropriate.

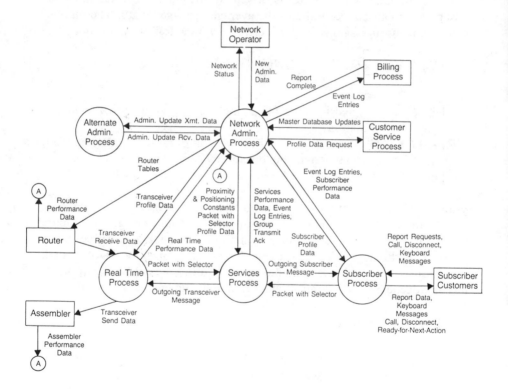

Figure 3.1 RDSS Central Software

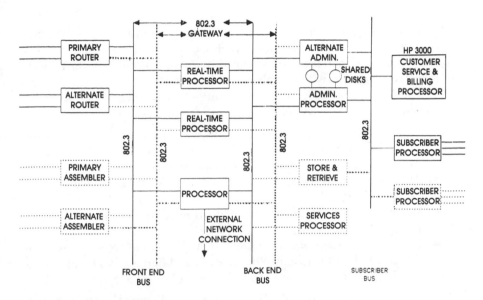

Figure 3.2 Central Hardware Configuration

3.1.4 System Management Issues for Other Satellite Systems Capable of RDSS Types of Services: Argos and Mobile Satellite Services

Argos and mobile satellite service concepts, such as the Inmarsat standard-C service, are capable of providing certain RDSS-type services, as described in Section 2.1.4. In this section, we shall examine some of the unique system management issues for these systems.

3.1.4.1 Argos

The capacity of Argos and like systems is defined as the maximum allowable number of data collection platforms (DCPs) that can transmit data through the system. For data collection systems aboard geostationary satellites (unlike Argos), the capacity is easy to assess, because

- All DCPs are permanently in view of the satellite;
- Each self-timed DCP transmits at a regularly assigned time;
- Each self-timed DCP is allocated a constant time slot for transmission;
- Each DCP is assigned a transmission frequency chosen from a fixed number of channels.

Transmission time and frequency are assigned in order to prevent interference, and hence to ensure relay of each message. The maximum number of allowable DCPs can then be calculated by simply multiplying the number of frequency channels by the number of time slots.

The Argos data collection and location system differs from a geostationary data-collection-only system in that

- At any given instant, only 3-4% of the earth's surface is viewed by the relatively low-earth-orbiting satellite, hence only a limited number of PTTs can be received by the satellite's receiving system;
- PTT transmission repetition periods are randomly distributed between 40 s and 200 s;
- Message duration is also randomly distributed, but between 0.36 s and 0.92 s;
- Transmission frequency is the same for all PTTs, but receiving frequencies are randomly distributed because of the difference in Doppler shifts associated with a random distribution of PTTs on the earth.

The main consequence of these differences in Doppler shifts is the risk of nonacquisition of a PTT message, due to either interference from simultaneous transmissions producing the same receiving frequency, or unavailability of the satellite receiving system. Prior to making a quantitative assessment of Argos-like system capacity, let us examine the satellite receiver payload in more detail.

The Argos payload aboard NOAA Tiros-N spacecraft is called the *data collection and location system* (DCLS). The fundamental parameters of DCLS are generic to low-earth-orbiting data collection systems. The DCLS receives the messages transmitted by PTTs within view of the satellite (referred to as the *visibility zone*). Because this is a random-access system (encoded PTT messages are received on a random basis), the received signal is a mixture of all the messages transmitted by the PTTs within the visibility zone. The DCLS attempts to accommodate a maximum number of messages, but there must be a limit to the number of processing channels available. Message separation in time is achieved by asynchronization of transmission and the use of different repetition periods, while separation in frequency is through the Doppler shifts of the various PTT carrier frequencies.

The DCLS consists of

- Power supplies;
- Processing equipment capable of handling a discrete number of messages at a time (provided that these are Doppler-separated in frequency);
- The receiving section, comprised of a receiver and a search unit, each being duplicated to ensure adequate redundancy.

At present, there are four sets of processing equipment aboard each Tiros-N spacecraft, each capable of processing one message at a time. Each such processing unit is comprised of

- A phase-locked loop for rapid frequency and phase synchronization on the unmodulated section of the message;
- A bit synchronizer, which generates a clock signal at the appropriate bit rate and also performs signal restoration;
- A Doppler counter for determination of receiving frequency;
- An encoder-formatter, which generates telemetry messages in a digital form (PTT ID number, sensor data, measured frequency, and time and date of measurement).

The Argos data flow is multiplexed with other data generated aboard the Tiros-N spacecraft and transmitted when in view of the telemetry stations. There is substantial similarity here with the PN generator identified in Section 2.2 for a dedicated RDSS satellite. However, an RDSS spacecraft can also be a pure "bent pipe," with no processing accomplished aboard the satellite.

With regard to system capacity we may define a *system use factor* as the mean rate of arrival of messages at the satellite receiver. From the user's viewpoint, system performance can be judged on the basis of two parameters:

- Elementary probability of message acquisition;
- Hit error probability in the sensor data part of the message.

These two parameters relate back to the system use factor. Various simulation tests of the Argos on-board package have provided useful information about message acquisition probability. For an acquisition probability of 0.8, Table 3.1 provides statistics on what the package can simultaneously handle. Note that the first column covers the late 1980s, the middle column covers the early 1990s, and the last column covers the mid-1990s. Growth is mostly due to incorporating additional parallel processors in future DCLS payloads.

Message acquisition probability is, of course, improved by repeating the message several times during the same pass. For example, for an elementary probability of message acquisition of 0.8, the probability of message reception becomes 0.9920 for a message repeated three times and 0.9999 for a message repeated six times. As to bit error probability, trials have confirmed the expected value of 10^{-4}; that is, one erroneous bit is encountered in 10,000. This percentage can be further reduced by ground processing operations when several identical messages are transmitted and received during the same pass. This improvement is possible because the processing software automatically compares all received messages.

The entire Argos data processing system is shown in the block diagram of Figure 3.3. Space segment to support this system is firmly scheduled through

Table 3.1

Number of Platforms Simultaneously in View and Able to be Collected

Platform	*Spacecraft*		
	NOAA-10 NOAA-D	NOAA-H NOAA-I NOAA-J	NOAA-K NOAA-L NOAA-M
Data Collection Platforms			
100 seconds; 256 bits	225	270	937
200 seconds; 256 bits	450	540	1875
200 seconds; 32 bits	1152	1382	4800
Location Platforms			
50 seconds; 256 bits	110	132	458
60 seconds; 256 bits	135	162	562
60 seconds; 32 bits	345	415	1442

Source: Service Argos, Inc.

---

at least 1995; NOAA-9 and NOAA-10 are now the operational satellites hosting an Argos DCLS, and six more with nominally two-year-lifetime NOAA spacecraft scheduled for launching, one per year from 1987 to 1993. Later NOAA spacecraft, those to be launched in the 1990s, will increase the system's capacity by a factor of four, due to additional parallel on-board processors.

3.1.4.2 Mobile Satellite and Inmarsat Standard-C Based Services

System management issues are somewhat unique for each type of mobile satellite service: land, aeronautical, and maritime. Only the maritime mobile satellite service is now in operation, and thus presents system management issues with regard to radiodetermination. There is still considerable uncertainty concerning other mobile satellite services. Clearly, system management issues will dominate system design if land-mobile and aeronautical mobile satellite services must share the same 1545-1560 and 1645-1660 MHz bands, as the Federal Communications Commission has mandated in the United States. The FCC has taken this position because these frequency bands have long been internationally allocated to the aeronautical mobile satellite service. The FCC has been faced with the facts of both the allocation and the aviation community's inability to use it for more than two decades, on the one hand, and the clear willingness of private companies to use promptly the allocation for land mobile satellite service, on the other hand. Thus, the FCC has required sharing

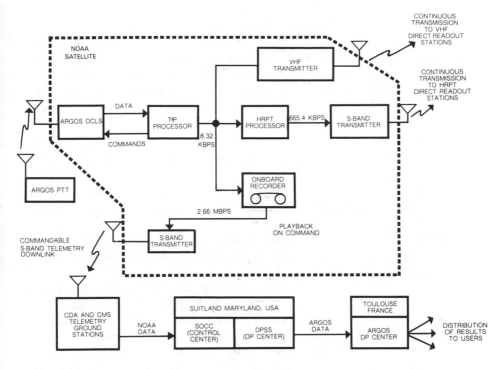

Figure 3.3 Diagram of Argos System Aboard NOAA Tiros-M
Source: Service Argos, Inc.

of the bands with priority of use rights for safety-of-life aeronautical satellite services.

The Inmarsat system, conversely, has a clear and unfettered right to operate throughout the 1530-1545 and 1626-1645 MHz bands. The key system management issues for this system involve the harmonization of a growing diversity of services. Although these services do not generally involve radio-determination techniques *per se*, they will be reviewed here because position reporting is a key component, and hence there is market overlap with RDSS. Also, it is important to remember that the purpose for which Inmarsat was founded in 1976 was "to make provision for the space segment necessary for improving maritime communications, thereby assisting in improving distress and safety of life at sea communications, efficiency and management of ships, maritime public correspondence services and *radiodetermination* capabilities" (emphasis added). This wording recognized that communication channels entail radiodetermination capabilities (as noted in Section 2.1.4), just as radiodetermination channels entail communication capabilities. Inmarsat's primary purpose, however, is communication. Hence, the system was designed as a frequency-channelized, narrow-band network.

The Inmarsat system is managed by a *secretariat* under the guidance of a *council*, representation on which is proportional to capital investment and use of the system. As of early 1987, the US COMSAT organization controlled approximately 30%, and the UK government an additional 21%, of the votes on the council. Management direction has been given to provide the following types of services to users in maritime regions:

- Voice and data (and occasionally video) services to and from relatively large and expensive standard-A shipboard earth stations (standard-A SES) capable of satellite tracking;
- Data services at 600 b/s to and from relatively small (15 pounds) and inexpensive ($10,000) standard-C SES, with nearly omnidirectional antennas;
- Network services under the rubric of *enhanced group call* (ECG), wherein messages may be broadcast to appropriately equipped ships by geographical area, ship ownership, or affiliation;
- Data services at 300 b/s, and eventually voice services, to aircraft with new terminal stations, which are still in the developmental stage;
- Position reporting services via either standard-A or standard-C SES;
- Narrow-band, tone-code ranging radiodetermination services;
- Satellite services related radio navigation, such as provision of a real-time GPS-Glonass integrity channel and transmission of differential correction data for GPS-Glonass.

In all cases, Inmarsat services are procured indirectly from national telecommunication administrations or registered private operating agencies (RPOAs), working from coastal earth stations established throughout the world.

Inmarsat's policy is that its system management will come to be increasingly oriented toward widespread use in ships, aircraft, and eventually land-mobile vehicles of the newly developed, data-only, standard-C terminal. This terminal tunes in 5 kHz increments throughout the maritime mobile satellite service allocation, except between 1626.5–1631.5 MHz. Various versions of the standard-C terminal permit either full-duplex or half-duplex operation. As Inmarsat's director of terminal development, J.C. Bell, states:

Standard-C may therefore become the workhorse of marine communications in the 21st century, just as the Morse key has been in the past. A particularly useful feature of the system is the ability to use any international alphabet or language a network can support when sending messages. In these highly competitive days, an important commercial benefit will be the ability to send sensitive information without others listening and reduce the "misunderstandings" which frequently occur from telephone calls. The ability of a standard-C automatically to receive Marine Safety Information sent through the Enhanced Group Call system will significantly reduce the workload of ships' personnel, while ensuring that all important messages are received, noted, and ... acted upon in a timely manner. For ships not requiring voice, standard-C may serve as the primary communications system.

Through an interface with a navigation receiver, or if Inmarsat implements general tone-code ranging, the standard-C terminal will enable a user to access various position-dependent Inmarsat services. EGC enables data in the shore-to-ship direction to be broadcast by geographical area of any size, as shown in Figure 3.4. The Safety Net EGC service, for example, will be used by administrations for the promulgation of NAVAREA and storm warnings, shore-to-ship distress alerts, and routine weather forecasts to the high seas and those coastal waters not covered by NAVTEX. A hurricane warning sent through the Atlantic Ocean region satellite could be addressed to an area in the Caribbean, covering the projected path of the tropical storm. A few hours later, an updated warning could be sent, addressing a revised area. At all times, only ships within the area specified by the meteorologist would display the message, ostensibly resulting in greater attention. Nevertheless, it then becomes imperative that all maritime vessels be properly equipped and that the Inmarsat positioning system be reliable. A similar application would involve a rescue coordination center receiving a distress call from a ship and using Safety Net EGC to request assistance from any ship within a circular area of any radius around the problem location.

Automatically broadcast NAVAREA reports will also be provided via Safety Net EGC, with the geographical regions equal to standard NAVAREA regions, as shown in Figure 3.4. In the future, NAVAREA broadcasts could incorporate corrections for on-board maritime data bases, including those for electronic charts, nautical charts, lists of lights, tide tables, sailing directions, and underwater shoals. It is fascinating to realize that if appropriately combined with a radio navigation receiver, an evolved capability from Safety Net could allow a ship to circumnavigate the globe continuously without *any* manual control. Integration and management of such capabilities provides the Inmarsat organization with significant challenges in the years to come.

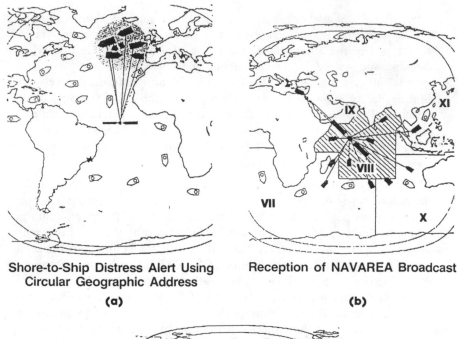

Shore-to-Ship Distress Alert Using
Circular Geographic Address

(a)

Reception of NAVAREA Broadcast

(b)

Local Area Warning Using
Rectangular Geographic Address

(c)

Figure 3.4 Inmarsat Safety Net Services
Source: International Maritime Satellite Organization (Inmarsat)

3.2 INTER-RDSS SYSTEM MANAGEMENT

There are three fundamentals to inter-RDSS system management. First, each system must avoid the use of spread-spectrum codes that lack *orthogonality* with regard to each other. Second, either a maximum power output on transceivers must be respected, or an average power-flux density limit per system must be imposed at the geostationary orbit. Third, a maximum power-flux density limit must be respected at the earth.

The use of pseudorandom noise codes by the RDSS system to produce the spread-spectrum modulation of the inbound link enables a large number of users to access the satellite simultaneously. The same process is also effective for extracting signals from interference generated by other RDSS systems in the same service area, provided that the signals are modulated by PN codes that have good cross-correlation characteristics. Gold codes, for example, are a class of binary codes that have such characteristics and they can be readily generated in large numbers. Gold codes are produced by the modulo-2 addition of two maximal-length codes. A maximal-length code is a binary code generated by a n-stage shift register, which has the maximum possible length for that size of shift register, 2^n-1 bits. [See R. Dixon, *Spread Spectrum Systems* (2nd Ed., John Wiley and Sons, 1984), pp. 58, 79.]

In the Geostar RDSS system, for example, Gold codes of $2^{17}-1$ chips in length are employed for the inbound link. There are 131,071 different Gold codes of this length available for each *preferred pair* of codes chosen from the 7710 maximal-length codes available. This means that there are, conservatively, in excess of 3.83×10^8 different Gold codes of length $2^{17}-1$ available to other RDSS licensees. Similar considerations prevail for the outbound link.

This method of coordination through the use of Gold codes permits an RDSS system to operate with other RDSS systems in the same area. For systems serving different but proximate coverage areas, sharing is further enhanced through antenna discrimination provided by a high-gain RDSS satellite antenna.

In the United States, the Federal Communications Commission established an industry committee of RDSS companies to establish intersystem management standards. The RDSS Technical Coordinating Committees *Standard No. 3* is that each RDSS operator in areas of overlapping coverage must demonstrate that "the partial correlation function between two systems, when evaluated over the data-bit period of the receiving system, should not exceed the negative of the spread-spectrum processing gain, plus 6 dB. In a system with a processing gain of 21 dB, for example, the partial correlation function evaluated over a data-bit period should not exceed −15 dB." This imposes little constraint because there are many thousands of codes with good cross-correlation characteristics from which to choose.

Coding alone, however, is not sufficient to ensure inter-RDSS system management if the RDSS system's average power levels are highly disparate. Therefore, adherence to PFD limits is needed in addition to the use of codes with good cross-correlation characteristics. The RDSS Technical Coordinating Committee's *Standard No. 1* is such that:

> The power-flux density in the 1610–1626.5 MHz band at the geosynchronous orbital arc produced by the aggregate emissions from all users of a single Radiodetermination Satellite Service system shall not exceed –155 dB watts per meter squared (in any 4 kHz band) for more than 0.01 percent of the time, and –158 dB watts per meter squared (in any 4 kHz band) for more than 50 percent of the time.

More simply stated is the RDSS Technical Coordinating Committee's *Standard No. 2*:

> The power-flux density at the earth's surface produced by emissions from a space station in the 2483.5–2500 MHz band for all conditions shall not exceed –139 dB watts per meter squared (in any 4 kHz band).

This power-flux density limit entails a maximum satellite EIRP of 56.6 dBW. Figure 3.5 plots level curves for PFD to show the extent of inter-RDSS system sharing possible before E_b/N_0 degrades below the requisite 10–12 dB level for adequate BER performance.

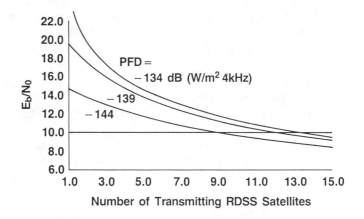

Figure 3.5 Power-Flux Density (PFD) Levels *versus* Bit Error Rate (BER) Limits
Source: FCC, Docket No. 84-690, Federal Communications Commission, Washington, DC, 1984.

For illustration purposes, let us work out a projected scenario quantitatively. Consider the case of four RDSS systems serving the same coverage area, each with three multibeam spacecraft, all of which transmit simultaneously. Such a scenario might be applicable to a region of continental size with separate aeronautical, governmental, and competitive private systems. The basic system characteristics are those of the ITU's advance-published USRDSS model, which is also used by the CCIR as its RDSS reference system.

For our working example, a user located on the –3 dB contour of the desired satellite could "see" one undesired adjacent beam *from the same system* of the desired satellite, 1 dB down, and two other adjacent beams, again, *from the same system*, 10 dB down. Further, the user could be in the main beams of the three undesired satellites from the three *other* RDSS systems. Table 3.2 shows the link analysis of this situation. As we can see, there is ample margin in the budget and the user will have no trouble communicating with the desired satellite. Further examples for other cases would show similar results. The reader should note that this analysis assumes co-polarization of all systems. If multiple RDSS systems are implemented with polarization diversity, then the total number of systems could be significantly increased, even doubled perhaps.

The inbound link presents a similar sharing picture, as shown in Table 3.3. In this example, it is assumed that the one wanted user is located at the 1-dB-down contour of the wanted antenna beam and that an average of 2.6 other users are accessing the same beam at a main-beam position. In addition to receiving power from the resulting 3.6 average users in the wanted beam area, the spacecraft antenna is receiving interference from RDSS users communicating with adjacent beam areas. Additionally, the table includes the interference effects of three other RDSS satellites from different systems, including both main-beam and adjacent beam contributions. The four eight-beam satellites represented in the table, derived from the CCIR's RDSS reference system, correspond to a total average user population of some 115 simultaneous users. If we assume a 20 ms message length, this corresponds to some 5700 users per second, or more than two million users per hour. Even with this user population, the wanted signal has an excess margin of greater than 3 dB. By working through this type of analysis, we acquire a good representation of inter-RDSS system management.

An important factor to keep in mind with respect to inter-RDSS system management involves network *homogeneity*. Therefore, the RDSS Technical Coordinating Committee adopted power-flux density limits at both the surface of the earth and the geostationary orbit. Compatibility becomes much more problematical with highly inhomogenous networks. Hence, advance coordination of RDSS networks is essential, and planning with regard to existing network parameters is wise.

Table 3.2
Outbound Link Analysis
(Satellite-to-User)

(a) *Parameters*	
Frequency	2491.5 GHz
Satellite power	19.5 dB (W)
Satellite gain	38.8 dB (i)
EOC correction	3.0 dB
Feed loss	0.7 dB
Satellite EOC EIRP	54.6 dB (W)
Atmospheric loss	0.7 dB
Path loss	191.8 dB
Obstacle margin	0 dB
Received power (isotropic)	137.9 dB
User G/T	−24.8 dB (1/K)
Boltzmann's constant	−228.6 dB (J/K)
C/N_0 (thermal)	65.9 dB (Hz)*
(C)	−134.9 dB (W)
*Satellite-to-user only.	

(b) *Adjacent Beam/Calculation (Relative to C)*		
No. Beams	*Level*	*Level*
1	0	−134.9 dBW
2	−10	−141.9 dBW
Adjacent beam noise entries		−134.1 dB

(c) *Other Systems (All Main Beam)*	
Number of systems	3.0
Relative power *versus* desired RDSS	−131.9 dBW
Noise level	−127.1 dBW
Total adjacent and other noise	−126.3 dBW
I (baseband contribution)	−195.5 dBW

(d) *Total Downlink*	
$(C/N_0 + 1°)$	59.5 dB (Hz)
Data rate 64000 b/s	48.1 dB (Hz)
Received Eb/N_0	11.4 dB
Required Eb/N_0	9.8 dB
Margin	1.6

Table 3.3
Inbound Link Analysis
(User-to-Satellite)

(a) *Parameters*	
Frequency	1618.3 MHz
User power	19.0 dB (W)
Feed loss	0.0 dB
Antenna gain	3.0 dB (i)
User EIRP	22.0 dB (W)
Atmosphere loss	0.7 dB
Path loss	189.0 dB
Received power (isotropic)	−167.7 dB (W)
Satellite G/T	3.0 dB (1/K)
EOC correction	−1.0 dB
Boltzmann's constant	−228.6 dB (J/K)
(b) *Additional Noise*	
CDMA NOISE CALCULATION	
Other users per beam	2.6
Relative level *versus* C (main beam)	1.0 dB
Noise power	−201.7 dB (W)
ADJACENT BEAM NOISE CALCULATION	
Number of adjacent beams	3.0
Relative level *versus* C (main beam)	−10.0
Number of users per beam (assumed)	3.6
Noise power	−206.5 dB (W)
OTHER SYSTEM INTERFERENCE	
Number of other systems	3.0
Main beam	−195.5
Adjacent beam	−201.7
Noise level	−194.5 dB (W)
Total self and other noise	−193.5 dB (W)
(c) *Total Uplink*	
$C/(N_0 + I_0)$	55.1 dB (Hz)
Data rate (16,000 baud)	42.0 dB (b/s)
Received E_b/N_0	13.1 dB
Required E_b/N_0	10.0 dB
Extra margin	3.1 dB

3.3 CONTROLLING INTERFERENCE WITH NON-RDSS SYSTEMS

As a new service entering an old region of the frequency spectrum, questions were certain to arise regarding interference between RDSS and pre-existing non-RDSS systems. The analogy to a new office building going into a long-ago developed portion of downtown is well taken. In the process of zoning for or permitting the new office building, regulators must take into account the pre-existing interests. Sometimes, these pre-existing dwellers are simply vacated, which is essentially what the FCC did in the United States to fixed microwave links in the 2492 MHz band. Other times, the new developer is obliged to refurbish the old facades and build a glass and steel office building around the classic old brick buildings. As described below, this is what occurred between RDSS and radio astronomy at 1610 MHz. Most often, the developer realizes that whatever he accomplishes on his lot, he must still live with the crowding and pollution of the downtown area. This is analogous to the requirement that RDSS ought not complain of interference from microwave ovens operating at 2450 MHz.

3.3.1 Sharing with the Fixed Microwave Service

Annex I to CCIR *Document 8/552-E* notes that due to the omnidirectional characteristic of the RDSS user-receiving antenna, the users are susceptible to interference from nearby fixed and mobile transmissions. In fact, the potential for interference is quite high if fixed or mobile transmitters are within the line of sight of the user's receivers. However, the intrinsically mobile nature of RDSS, and its acknowledgment-based protocol, should minimize the operational effect of such interference.

Nonetheless, the -139 dB (W/m^2) PFD of the outbound downlink, referenced to a 4 kHz bandwidth, exceeds the CCIR recommendation for protection of fixed microwave systems from satellite interference. The CCIR recommendation for the 1.7–2.5 GHz band is that satellite PFD be restricted to a range of -154 to -144 dB (W/m^2) in 4 kHz, varying with the angle of arrival of the emission. However, these limits were developed on the basis of protection for a hypothetical 2500 km, 60-hop, radio-relay system. Almost without exception, worldwide fixed microwave use of this band is limited to short, one or two hop systems for which -134 dB (W/m^2) in 4 kHz would provide protection equivalent to that recommended by the CCIR. The -139 dB(W) figure adopted by the RDSS Technical Coordinating Committee appears to be a reasonable compromise between protection of existing systems and ensuring adequate margin for multiple RDSS systems. The FCC, however, has made the rights of fixed microwave users in the 2492 MHz band secondary to those of RDSS operators.

3.3.2 Sharing with Other Satellite Services

No other satellite services are allocated bandwidth within the RDSS allocation, although such services do enjoy allocations immediately adjacent to the 1618 MHz band. Immediately above the RDSS band is the maritime mobile satellite service, operated by Inmarsat, occupying the 1626.5–1640 MHz band for its uplink. Immediately below the RDSS band, at 1570–1610 MHz, is the radio navigation satellite service, within which both the US and the USSR operate military broadcast-like positioning systems. Out-of-band services, such as Inmarsat and the US Air Force's Navstar system, are protected by an emission-limitation rule adopted by the FCC.

The emission-limitation rule, governing RDSS operations in both L and S bands, provides that emissions at a frequency removed by more than 50% from the assigned frequency shall be attenuated below the mean power density at the assigned center frequency, as specified in the equation: $A = 12 + 0.2(P{-}50)$, where A = attenuation in dB below the mean power density level, subject to a maximum of 75 dB, and P = percent of assigned bandwidth removed from the carrier frequency. This formula adequately protects Inmarsat and other systems. It is interesting to note that the Soviet and US radio navigation satellite services utilize spread-spectrum modulation, and hence are particularly immune to interference. Indeed, they share a common band, and each system employs some 24 simultaneously transmitting satellites.

3.3.3 Sharing with Radio Astronomy

Radio astronomers use the 1606–1613.9 MHz band for observations of a hydroxyl line. Due to the extreme sensitivities required by radio astronomy receivers, it is not possible to perform observations of the hydroxyl line when a transmitting RDSS user is within the line of sight of an observatory. In order to protect these valuable observations, a plan has been developed for coordinating RDSS and radio astronomy usage of the 1610 MHz region.

The radio astronomy and RDSS sharing plan has been adopted as law in the United States. The law provides as follows:

(1) All Radiodetermination Satellite Service licensees will automatically restrict user transmissions to occur only within the first 200 millisseconds following the one-second time marks of Coordinated Universal Time when users enter Radio Astronomy Regions during a period of radio astronomy observations in the 1606.8–1613.8 MHz band. Any segment of a Radio Astronomy Region that is part of a Consolidated Metropolitan Statistical Area is not subject to coordination and transmission restriction limitations.

(2) Each Radiodetermination Satellite Service licensee will establish an observation notification procedure through the Electromagnetic Spectrum Management Unit, National Science Foundation, Washington, D.C. 20550, that satisfactorily provides for the restriction of user transmissions as described above during periods of radio astronomy observations in the frequency band 1606.8–1613.8 MHz.

Radio astronomy regions are defined as a region centered on certain major observatories with a radius of 150 km in the air and 25 km on the ground. An effect of these coordination restrictions is to increase slightly the response time to user transmissions because the transmissions can occur only within certain periods of time. A second effect is to increase the integration time for radio astronomy observations of the hydroxyl line. These restrictions, however, permit two apparently incompatible services to utilize the same frequency band with acceptable operational complexities.

A further problem between RDSS and radio astronomy involves the spectral distribution characteristics of spread-spectrum transmissions. Spread-spectrum modulated transmissions, by their very nature, occupy an expansive amount of bandwidth at very low power densities. This causes no problems for most radio services, but it is highly pollutive toward the ultrasensitive radio astronomy service.

The unfiltered output of a transmitter with square-wave phase modulation has a very broad spectrum, described by $P_f = (\sin x / x)^2$, where $x = \pi(f-f_0)/f_C$, f_0 is the center frequency and f_C is the chip rate of the modulation. In the particular case of a standard 8 Mcps rate, the sidebands would fall to an acceptable level of 38 dB below the peak power density of the transmitted signal at a frequency separation of 200 MHz from the center frequency. This very large separation would cause problems in a number of different radio astronomy bands, including the treasured 1400–1427 MHz band. Further distressing to the radio astronomy community is that such distant out-of-band interference can be caused at a separation of hundreds of kilometers if the RDSS transceiver is airborne. Figure 3.6 depicts the RDSS spectral mask. Filter performance needed to reduce out-of-band emissions to acceptable levels is shown in Figure 3.7, with the new spectral mask depicted in Figure 3.8.

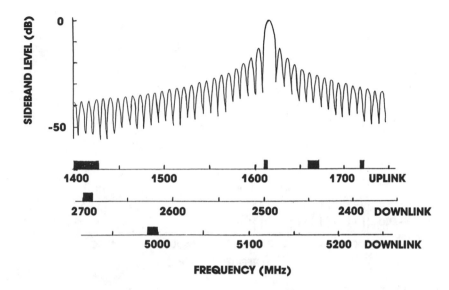

Figure 3.6 Unfiltered Spectrum
Source: Committee on Radio Frequencies, National Academy of Sciences

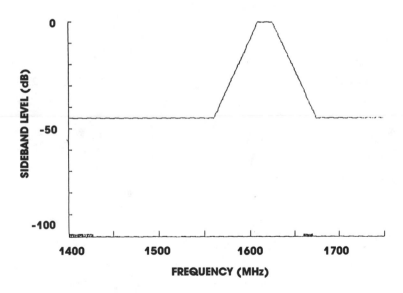

Figure 3.7 Filter Response
Source: Committee on Radio Frequencies, National Academy of Sciences

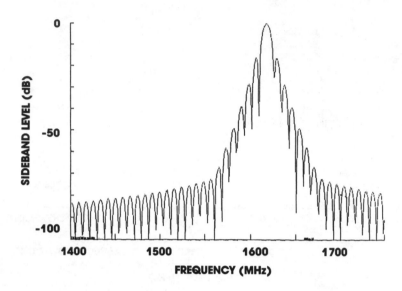

Figure 3.8 Filtered Spectrum
Source: Committee on Radio Frequencies, National Academy of Sciences

Chapter 4
System Applications

RDSS appears capable of becoming one of society's most widely used radio communication systems. The purpose of this chapter is to explicate specific RDSS applications in aeronautical, maritime, land-mobile, personal, and specialized markets. By necessity, and limited imagination, we have omitted many potential applications.

4.1 AERONAUTICAL RDSS

There is no market in which RDSS can make a more unique contribution than that of aviation and aeronautics. Communication and navigation, the hallmarks of RDSS, are fundamental to safety of flight. This section provides a precis on current aviation issues in these areas and the contribution of RDSS technology as a systemic solution.

To place the issue in context, it is fascinating to reflect on the fact that each day in the United States alone approximately one million people are airborne. Similarly amazing is the fact that on an average day as many as 10,000 aircraft will be airborne over North American skies.

4.1.1 Communication, Navigation, and Surveillance (CNS)

The term *communication, navigation, and surveillance*, or CNS, is the buzz word of aviation experts looking at the next generation of aviation electronics. *Communication* refers to the ability of the aircraft to remain in two-way contact with ground controllers. While communication is almost always available in the developed world, it is generally lacking in areas such as Africa and the maritime realms. *Navigation* is the ability of a flight crew to determine their position and set vectors to a desired location. Large aircraft have expensive navigation systems that work anywhere in the world. The very low frequency Omega system provides navigation signals, of relatively low accuracy, worldwide. Inertial navigation systems can also be relied upon, but the navigation equipment is very limited on small aircraft. Paper maps, a compass, and in recent times a Loran-C receiver are normally the extent of a small plane's navigational aides. *Surveillance* refers to the ability of ground controllers to know the location of aircraft in the sky. At present this is accomplished through ground-based radars, reporting altimeters, and crew reports.

RDSS is an integrated CNS system. Two-way digital communication replaces the awkward VHF radio system currently in place. Navigational information is automatically provided every time the pilot accesses the system. This navigational information may include data regarding obstacles, such as radio towers or mountain ridges, as well as safety aides including emergency landing strips. The control center enjoys complete surveillance information on all aircraft. An RDSS system is especially valuable for CNS in developing countries, where little such infrastructure exists.

4.1.2 Collision-Avoidance System Technology (CAS)

A *collision-avoidance system* (CAS) can be defined as a "blunder-resolver," which will only be used when something goes wrong with planned conflict-free flow of air traffic, or if there has been some unexpected and unsafe infringement on safe, scheduled, air traffic flow. An RDSS-based CAS can present each equipped aircraft with a 30–60 nmi (ATC) situation map with one's own position being close to the center of the display. The map can also show runways, navigational aids, fixes, air traffic control glideslope, and intended flow of all close-proximity aircraft. The intended flows will, if applicable, show at least one maneuver point of each aircraft. The intrinsic value of displaying both intended and actual flow is that a pilot can select an escape maneuver based on the flow of all aircraft in the plane's immediate vicinity.

When the National Airspace System ATC was developed, there existed no centralized surveillance system. This fact may explain why each air route traffic control center (ARTCC) was treated as an independent entity. Each ARTCC only addresses the flow into and out of its terminal area. No effort is made either to define or maintain a conflict-free flow for any user of the entire airspace. RDSS, on the other hand, has the potential to be a primary source of CNS for ATC and could well be the basis for a centralized CAS. In the control center, a centralized surveillance system of the entire airspace can exist. The ground-based computer could be constantly checking for conflicting air traffic flows and determining the control action required to prevent the need to exercise CAS logic. Individual aircraft, with data sent to them from the control center, can perform similar functions for their immediate vicinity.

In an ultimate manifestation, we can imagine a completely automated aircraft flight and air traffic flow control system built around RDSS. For example, a transceiver emission could be generated upon turning on the ignition switch of the aircraft. The control center would know the type of aircraft (based on subscriber registration information) and the airport at which it is located (from RDSS ranging data). The control center would draw from memory the necessary statistics on the airport's taxiways and runways, and could access local weather information as well. The passenger could then key in a desired location such as "LAX" (Los Angeles International Airport). After running through a checklist, with some passenger participation, the plane could be automatically flown to its destination by control center programs. With complete knowledge of all aircraft in use and relevant terrain features, hundreds of thousands of light aircraft could be safely flown via RDSS links and computational capabilities.

4.1.3 Alternative Approach to the Emergency Locating Transmitter (ELT) Problem

All aircraft are required by law to be equipped with an *emergency locating transmitter* (ELT), designed to transmit on a 121.5 MHz frequency upon aircraft impact with the ground. High flying aircraft and special satellites stay tuned to this frequency in hope of detecting a transmission. Implementation of this system has helped significantly in finding downed aircraft and getting quicker help to any survivors.

There are two fundamental problems with the ELT system. First, more than 98% of the ELT transmissions detected are false alarms, triggered by a hard landing or other jolt to the ELT. This results in a tremendous waste of rescue resources. Also, such false alarms have been known to mask true emergency transmissions! Second, there are many instances in which the ELT does not transmit upon a severe impact and, when it does, accuracy is usually not better than 10 miles. Both of these problems emanate from the concept that the ELT is carried in an aircraft as dead weight at all times until a crash, at which time it is supposed to work.

RDSS offers an alternative approach to the ELT concept. With RDSS, the control center's computer monitors the aircraft's position at all times in flight (surveillance). The computer compares this flight path with a stored digital terrain map that includes airfields, obstacles to aeronautical navigation, and flat areas suitable for landing. When the computer detects that the plane is on a track that is intercepting terrain, away from an airport, the control center sends "mayday" messages to the aircraft. If the aircraft does in fact crash, the control center knows the precise location of its demise.

4.1.4 Aeronautical RDSS Implementation Considerations

It is likely that an aeronautical RDSS system would be designed, built, and operated exclusively for aeronautical purposes. Here, we review system capacity, regulatory process, and economic considerations for implementing such a system.

The customary metric for assessing system capacity requirements is an air traffic control region's *peak instantaneous aircraft count* (PIAC). Table 4.1 provides PIAC estimates developed by the Radio Technical Commission for Aeronautics (RTCA) in its Report on User Requirements for Future Communications, Navigation and Surveillance Systems, including Space Technology Applications (1986).

Table 4.1

RTCA Estimates of Peak Instantaneous Airborne Counts (PIAC) for Various Geographic Areas (1995–2010)

Area	PIAC
Continental US	50,000
Busy en route center	5,000
Busy terminal center	1,500
Canada	4,000
Mexico, Central America	600
South America	2,600
North Atlantic	800
Caribbean	400
South Atlantic	150
Pacific Basin	1,200
Europe	2,200
Africa	2,200

Aircraft are normally distributed among terminal areas (about 20%), near terminal areas (about 20%), and en route (about 60%). With an RDSS system, the aircraft position-fix rate can be adjusted in accordance with phase of flight, which will be known to the control center by virtue of the data it receives. Average requirements turn out to be a position update about every 10 seconds, based upon:

- 90 min average flight durations in the United States;
- 50 min average cruise, with fix interval of 20s;
- 20 min average climb, with fix interval of 10s;
- 20 min average descent, with fix interval varying from 10 s at 600 m, to 5 s at 300 m, to 2 s at 150 m, to 0.6 s at landing.

With a position fix on average every 10 seconds, the RDSS system capacity requirements range from 1000 transmissions per second for the current US PIAC of 10,000 aircraft to 5000 transmissions per second for the projected 21st century PIAC of 50,000 aircraft. Both figures are well within the probable capacity of a multibeam RDSS system. RDSS systems are limited to the capacity of their outbound channel, nominally 64 kb/s in the CCIR reference design. If we assume the average length of a control center-to-aircraft transmission is 256 bits, each RDSS outbound beam handles about 250 transmissions per second. This implies a current need for a four-beam satellite, with growth to perhaps a 20-beam satellite in the early 21st century.

The regulatory process for aeronautical RDSS involves industry committees, national aviation authorities, and, in many instances, national legislative bodies. In the United States, the Radio Technical Commission for Aeronautics (RTCA) is the industry committee that recommends new standards and regulations to governmental authorities. In 1986 the RTCA established a new special committee to develop minimum operational performance standards for RDSS. Once agreed to, these will likely be submitted to the Federal Aviation Administration (FAA) for action. If the FAA approves the RTCA recommendations, it will establish a technical standard order (TSO) incorporating the RTCA recommendation as is, or with modifications. Once a TSO is in place, aircraft manufacturers can obtain type certifications for RDSS electronics integrated into aircraft avionics systems. If the FAA process becomes inefficient, RDSS providers have the option of going directly to Congress for authority to integrate RDSS into the nation's avionics inventory. It was Congress, for example, which mandated the use of ELTs over the objections of the FAA.

Economically, somebody must pay for aeronautical RDSS. The possibilities are as varied as direct purchases by the aircraft owners, subsidization by the federal government, or perhaps even installation by aircraft insurers as a means of reducing aircraft-insurance underwriting losses resulting in particular from midair collisions. One attractive concept involves installation of an RDSS transceiver prior to approved take-off, and removal upon landing, clearly reducing the total number of transceivers to be made available for purchase. We should note that alternatives to mandatory carriage of an RDSS transceiver as a CAS tool are very expensive. The currently proposed CAS system in the US, TCAS-2, will involve at least $100,000 worth of additional electronics in each aircraft. It has also been noted that RDSS transmissions could be used by the government as a means of charging for airspace transit.

4.2 MARITIME RDSS

The maritime community was the first set of users to benefit from radio communication technology. Indeed, radiodetermination as a concept traces its ancestry back to maritime navigation requirements *circa* 1910. The new field of hydrography is in large part the result of maritime radiodetermination technology. Hydrography encompasses the study, description, and mapping of oceans, lakes, and rivers, especially with reference to their navigational and commercial uses. Hydrographic knowledge is of fundamental importance to modern maritime operations as diverse as safe vessel navigation, offshore resource exploitation, and monitoring the movement of water-borne pollutants.

4.2.1 Maritime Requirements for RDSS

In the United States, the Radio Technical Commission for Maritime has established a special committee, known as SC-108, to develop minimum operational performance requirements for maritime RDSS transceivers and systems. The committee, which represents both commercial and governmental maritime operators, developed the following list of user applications for RDSS in 1986:

- Improved emergency position-indicating radio beacon (EPIRB);
- Vessel traffic services position information and reporting;
- Canal-waterway lock control;
- Anchorage monitoring and control;
- Boat dispatching;
- Fishing area identification;
- Yacht race surveillance;
- Ocean dumping monitoring and control;
- Commercial fleet control and dispatching;
- Surveying, both seismic and hydrographic;
- Dredging boundary identification;
- Establishment and replacement of aids to navigation (navigational buoys);
- Monitoring the location of aids to navigation;
- Search and rescue planning control;
- Rendezvous of ships;
- Distress location;
- ETA-logistics support arrangements;
- Treasure hunting;
- Weather balloon tracking;
- Monitoring and control of traffic-separation schemes;

- Silent alarm for tracking suspect or stolen boats;
- Fishing vessel treaty compliance reporting;
- Lifejacket and lifeboat beacon;
- Rig positioning;
- Hazardous or restricted navigational area identification;
- Way-point navigation;
- Oil spill tracking.

In general, hydrographic surveying tends to impose the most stringent accuracy requirements. The US Federal Radio Navigation Plan has defined the requirement as variable from 1–5 m (harbor and harbor-approach areas) to 10–100 m (oceanic regions), for both predictable and repeatable accuracy, all 2 drm (direction of relative movement). By contrast, accuracy requirements for safe navigation to preclude grounding or collision range from 8–20 m (harbor approach) to 2–4 nmi (oceanic regions). User needs that share such a high order of accuracy requirement include marine scientific research, commercial fishing, petroleum or mineral exploration, and certain naval operations. Seismic surveying in the resource exploitation industry probably has the most rigid requirement with a repeatable accuracy on the order of 1–5 m, and a fix rate of once per second.

A particularly difficult requirement involves the 40,000 navigational buoys and other floating navigational aids maintained by the US Coast Guard. Each such aid must be initially positioned within a precise point determined by an appropriate chart. The chart itself is tied to a local datum, such as the global geodetic datum WGS-84. The positioning aids often must maintain accuracy of positioning to under 10 m in order to support the Federal Radio Navigation Plan navigational accuracy requirement of 20 m for large ships in harbor areas, all 2 drm. Hence, maritime authorities need reliable, continuous verification that an aid is keeping station within limits, an alarm when said limits are exceeded, and a means to notify users of the discrepancy until accomplishment of relocation. We should recall that each year numerous ships are lost to collisions and groundings, some of which are because of mislocated navigational aids, with severe environmental, safety, and economic consequences.

There is also a requirement to determine and mark positions on the ocean floor for a number of purposes. With the international adoption of *exclusive economic zones*, resource management of the ocean will give rise to boundary demarcation, law enforcement, and industrial mining rights. Proper management of the sea floor requires the establishment of marine benchmarks, or "monuments," to define jurisdiction. Acoustic ranging is the primary means for determining the sea floor positions. The acoustically derived relative position is then matched to a geodetic position through a navigation system onboard the monument vessel.

A final set of important requirements are those for preparation of general navigational charts. General purpose charts are those used by mariners for planning voyages, tracking their progress, and a wide variety of maritime business activities. The International Hydrographic Bureau (IHB) is principally concerned with the preparation of these charts to support safe navigational requirements. For example, shipping economics tend to favor larger vessels with deeper drafts. Thus, there is a need for very detailed surveys of harbor approaches and the continental shelf to a depth of about 40 m.

4.2.2 RDSS Applications to Hydrographic and Other Maritime Needs

RDSS appears well suited for medium-accuracy surveying work, monitoring floating aids to navigation, and performing automated, continuous surveying work. With regard to harbor-approach guidance, based on IHB criteria, RDSS may not be sufficiently accurate. High accuracy, short-range radar systems appear to make the most sense for the harbor environment. Nevertheless, RDSS may offer an important supplemental capability.

RDSS can be extremely valuable in the offshore environment, where it should be possible to simplify greatly the completion of surveying work in most coastal and oceanic areas and to capture this surveying data with high precision and low cost. Surveying work will be greatly facilitated by the simplicity of user interface with RDSS, the fact that all calibration is performed by the control center for the entire system, and the fact that RDSS systems would ordinarily be tied into the WGS coordinate frame. Also of significance to hydrographers is the fact that RDSS is continuously operating, requires no set-up time, and with its two-way messaging capability enables surveying data to be transmitted in real time, which permits shore-based hydrographers to redirect a surveying effort already afloat immediately to areas identified to be of special interest. With current procedures, which are dependent on postcruise processing of surveying data, it is very costly to repeat the survey's conditions when "holes" in the data develop.

RDSS is also well suited for the above-mentioned buoy positioning and tracking requirements. For example, RDSS can be used for the initial installation of a floating aid, well within the accuracy limits specified by the IHB. RDSS also permits periodic position reports, coupled with any status reports that are needed by the buoy monitoring center, such as the health of the aid's electrical power system. When the monitoring center senses that the position of the aid is out of watch-circle tolerance limits, it can order the aid to display a new or different audiovisual signal as a warning to mariners of its defective position. This alternative display gives "breathing time" to the responsible maritime safety authority for relocating the aid to its charted position, again with the assistance of RDSS.

One of the more subtle, but no less significant, potentials of RDSS is to transform the entire work of hydrography from a series of largely independent, time-consuming activities (survey, chart preparation, distribution, and correction) into an integrated process that involves both mariners and hydrographers. Instead of labor-intensive shipborne surveys, future work may be done by automatic survey vessels fitted with precision fathometers that are directly tied to the RDSS transceiver. On the mariner's side, the availability of RDSS could be merged with electronic charts. Mariners wishing to have the latest chart-correction data could directly access hydrographic charting centers via RDSS and have their shipboard data base updated via the outbound link. Thus, there is some promise for a closed-loop, RDSS-based hydrographic system, with integrated provision of all essential elements from initial survey to the preparation and maintenance of electronic charts.

4.2.3 Future Global Maritime Distress and Safety System (FGMDSS)

Existence of numerous incompatible radio systems aboard ships has long been a safety and operational hazard in the maritime community. Under the auspices of the International Maritime Organization (IMO), the international community conceived in late 1979 of a *Future Global Maritime Distress and Safety System* (FGMDSS), whereby all ships could communicate with each other and coastal stations. Progress in implementing FGMDSS has been slow, largely due to the reluctance of many seafaring countries to impose often costly international regulatory requirements upon their shipowners. FGMDSS radio communication equipment can cost many tens of thousands of dollars.

RDSS may well be the most effective way to bring into being a true FGMDSS. The low cost of an RDSS transceiver and its global range with appropriate satellite relays make it amenable to universal adoption. RDSS also functions as a highly effective navigational aid. RDSS can be used to navigate, regardless of visibility conditions. Warnings of impending collisions with other sea vessels can be given via RDSS. A digital map of navigational obstacles such as shoals can be readily made available to any seafarer. Even purely recreational activities can be accommodated, for example, returning to the precise location of a favored fishing spot.

Maritime RDSS is, virtually by definition, an international endeavor. By international treaty, the provision of maritime mobile satellite services has been reserved for the Inmarsat organization. The question is whether Inmarsat is also the appropriate entity to provide RDSS, or how it might be provided otherwise.

The Inmarsat treaty references radiodetermination as a service that the organization might provide, but not one which is among its primary purposes. The distinction is important because no other signatory to the Inmarsat treaty may operate a service which significantly competes with a service that is a primary purpose of the organization. In 1986 Inmarsat announced for the first time that it was interested in providing RDSS.

It is conceivable that other entities might undertake to provide RDSS over oceanic areas, but find it difficult to justify the costs of such activity. In Inmarsat's fifth year of service it obtained its 5000th customer, which reflects the low traffic volume in oceanic areas. More likely would be national or regional RDSS systems providing coverage in heavily traveled coastal waters and leaving RDSS service in the high seas for an international organization.

4.3 LAND-MOBILE RDSS

The land transportation market consists of those persons, companies, and organizations who need telecommunication and position determination services for the trucks, automobiles, and railway cars that they operate. Included in this market are the following land-mobile activities:

- Long-haul trucking, both common carrier and private;
- Local area delivery, including express delivery;
- Railway operations;
- Corporate and organizational automobile fleets.

RDSS can meet the needs of this market by providing drivers, fleet dispatchers, shippers, and buyers with an easy to use and economical means of contacting vehicles while they are en route and, at any time, ascertaining the geographical location of such vehicles. With an RDSS transceiver, a user in the land transportation market can do the following:

- Send messages such as new pick-up or delivery points to, and receive such messages from, any other person equipped with a transceiver, regardless of whether they are nearby or far away;
- Provide mills, factories, and assembly plants that rely upon "just in time" inventory-maintenance techniques with a means of continually monitoring the progress of important supplies;
- Offer fleet managers a highly efficient tool for increasing "backhauling," and thereby reducing "deadheading" (shipping cargo to terminal points from which there is no new cargo to be loaded for distribution to other points), one of the most costly problems faced by the trucking industry;
- Permit shippers, railway, and trucking companies to manage their in-transit assets in real time;
- Reduce theft and hijacking by automatic alerts of any crime in progress to the nearest law enforcement authorities;
- Give dispatchers a reliable and economical means of controlling their dispersed fleets and using them as efficiently as possible.

4.3.1 Land-Mobile RDSS Market Description

The communication, positioning, and radiolocation needs of the land transportation industry are not being satisfied at the present time. Communication service is currently limited to discrete, single urban areas through radiotelephone common carriers or private systems and therefore is of little utility to intercity and interstate trucking. Cellular telephone systems enable long-distance mobile communication service from major urban areas, but at high cost and without the ability either to make or receive calls when between such areas. Public radiotelephone service, including cellular, is being provided to about one million subscribers in the United States by several hundred, mostly small, radio common carriers and large telephone company subsidiaries.

Most radiotelephone service is provided by *private* systems. Latest FCC figures set the number of two-way land-mobile radio stations in these systems at nearly 1.5 million. As with the public radiotelephone service, however, these systems are strictly limited to a small local coverage area. Operational and maintenance costs of these systems are on a par with the costs of publicly available service.

Positioning and radiolocation services are not generally available for land transportation vehicles. However, based on interviews with numerous, representative trucking industry firms, a NASA report estimates that most of the 1.5 million long-haul trucks that are actively used would be in the market for such a positioning and voice communication service at an appropriate price (see Figure 4.1). The purpose of such a service would be to transmit a truck's location automatically to supervisory headquarters or the shipper's delivery points to monitor and track cargo containers as well as to pinpoint for law enforcement authorities the location of any truck hijacking or theft.

4.3.2 Land-Mobile RDSS Market Potential

The market potential for land transportation RDSS may be divided into long-haul and local area components. The long-haul component encompasses interstate trucking and railway operations, while the local area component includes existing public and private users of two-way radio service.

4.3.2.1 Long-Haul Submarket

The number of trucks operating on a long-haul basis has been growing by about 2% per year since 1977. By 1990 there will be more than 300,000 trucks in long-haul-for-hire transport and at least one million trucks in long-haul private transport. About half of the for-hire trucks will be in common carrier service and about half in long-term contract carriage. Hence, the

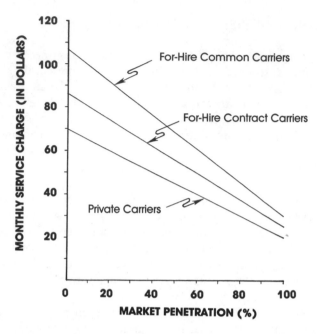

Figure 4.1 Estimated Price Sensitivity of Voice Communication in the Inter-city Trucking Industry

potential RDSS long-haul land-mobile market may be set at the projected total of for-hire and private intercity trucks, which is approximately 1.5 million during the 1990s.

A separate RDSS submarket could be defined for high-value cargo containers or trailers carried by trucks and railways. There are about 2.5 trailers per intercity truck and about 1.8 million railway cars. About 10% of this number represents high-value cargo or dangerous materials. Among the functions RDSS can provide to the owners and haulers of such containers include mobile inventory management, crime prevention, and meeting regulatory requirements for hazardous chemicals and refrigerated biological items. In summary, a total potential RDSS land-mobile, long-haul market could be defined for the 1990s as approximately 1.5 million appropriate trucks and more than 400,000 containers of high-value goods or hazardous substances.

4.3.2.2 Local Area Submarket

The best indication of the potential and diversity of the RDSS local area market comes from the lists developed by manufacturers (Motorola and General Electric) of historical and projected land-mobile radio terminals. Table 4.2 provides the tabulations officially submitted to the FCC by these firms for

Table 4.2
Licensed Land-Mobile Radio Transmitters in Use

(a) *GENERAL ELECTRIC CO. ESTIMATES (in thousands)*					
	1980	*1985*	*1990*	*1995*	*2000*
Police	1,216	1,479	1,803	2,188	2,664
Fire	346	403	464	538	626
Local government	416	556	744	998	1,334
Highway maintenance	144	184	236	302	385
Forest conservation	166	210	271	346	442
Special emergency	368	608	814	945	1,094
Total: public safety	2,656	3,440	4,332	5,217	6,545
Special industrial	420	486	564	656	757
Business	2,336	3,595	5,530	8,509	13,094
Power	591	788	1,059	1,418	1,894
Petroleum	140	162	188	219	254
Manufacturing	109	140	179	228	289
Forest production	46	50	57	61	70
Industrial radiolocation	26	42	68	109	175
Motion picture	6	6	7	7	8
Relay press	9	10	10	11	11
Telephone maintenance	140	206	320	446	652
Total: industry	3,817	5,485	7,964	11,664	17,204
Railroad	1,208	1,334	1,470	1,628	1,794
Taxi	127	149	171	197	228
Automobile emergency	21	23	25	28	32
Motor carrier	188	241	306	389	499
Total: transportation	1,544	1,747	1,972	2,242	2,503
Total	8,017 *	10,672	14,268	19,123	26,252

(b) *MOTOROLA CO. ESTIMATES (in thousands)*			
	Public Safety	*Industrial*	*Land Transportation*	*Total*
---	---	---	---	---
1958	304	417	201	942
1960	372	640	258	1,271
1965	545	1,316	416	2,277
1970	779	2,018	573	3,371
1975	1,072	2,830	680	4,582
1980	1,428	3,859	781	6,068*
1985	1,840	5,116	871	7,827
1990	2,315	6,623	952	9,890
1995	2,844	8,373	1,021	12,238
2000	3,411	10,384	1,084	14,879

*It is interesting to note that in 1972 the Electronics Industry Association (EIA) estimated that in 1980 the number of *licensed* transmitters would be seven million.

its *Final Report on Land-Mobile Radio Requirements* (1986). This same report concluded that there was a significant historical correlation of growth in land-mobile radio terminals with that of computers and peripheral equipment, in addition to less important correlations with other factors (see Table 4.3). On the basis of their research, the FCC staff developed a model to predict growth in land-mobile transmitters under a ratio of eight user terminals per station or transmitter. The results of the FCC model, and its good fit with General Electric and Motorola forecasts, is provided in Figure 4.2.

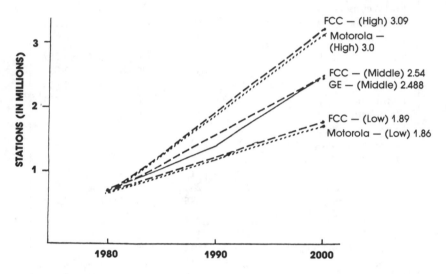

Figure 4.2 National Projections of Total Stations in the Private Land-Mobile Radio Services
Sources: From projections of FCC staff, General Electric Co., and Motorola Co. (1985–2000)

4.3.3 Land-Mobile RDSS Competition

No competition to RDSS in the trucking portion of the long-haul market appears on the horizon. This is because such a service cannot be provided economically with a terrestrial system, and that RDSS are the only satellite-type systems scheduled for launch. Railroad companies do have separate communications and crude positioning technologies available to them. Most large railroads operate their own private microwave links along their rights-of-way, and many have now procured computerized optical scanning systems

Table 4.3

Growth in Land-Mobile Stations, Communication, and Computer
Equipment (Constant Dollars, US$)

Type	Change (%)	Average Annual Rate (%)
1959–1970		
Land-mobile stations	253	12.2
Communication	123	7.6
Computers and peripheral equipment	253	12.2
1971–1980		
Land-mobile stations	254	15.1
Communication	107	8.4
Computers and peripheral equipment	192	12.6

Source: FCC, Final Report on Land-Mobile Radio Requirements, Federal Communications Commission, Washington, DC, 1986.

for keeping track of their cars. The microwave systems are generally not practical over long distances, leaving much trackage "dark," and the scanning systems are highly failure prone. To the railroad industry, RDSS may be more important as an industry-wide signalling system than as a car-locating system. One company, Railstar Control Technology, is currently marketing an RDSS-based railway signalling and control system.

With regard to two-way local area applications, it is clear that cellular radio constitutes the primary source of competition to RDSS. Traditional two-way radio systems, both private and public, are an additional choice for government-licensed purchasers. Some of these systems permit two-way data transfer between portable terminals that may resemble RDSS terminals. The difference is that RDSS permits nationwide coverage and a switched service.

RDSS also enjoys an important cost advantage over local area radio systems. The high costs of constructing and maintaining cellular and advanced computer-controlled mobile communication systems is expected to keep radiotelephone service costs well over $100 per month on average. This amount is three to 10 times higher for long-distance-connected intercity service (which is possible only with cellular). RDSS costs, based on required discounted cash flow to demonstrate a 30–50% return on investment, can range from $10 to

$50 per month. Of course, RDSS does not support full-duplex voice service, which may imply that radiotelephone and RDSS are in fact complementary. One example of this is the ability of RDSS to enable cellular telephone calls sent out to roaming cellular users. A roamer cannot be called without *a priori* location knowledge. RDSS provides this knowledge at little incremental expense.

4.3.4 Land-Mobile RDSS Market Share: The Voice Issue

There are two major factors that limit the portion of the potential market that RDSS can capture: (a) it is *nonvoice*; (b) it is *nonprivate*. It is doubtful that there is a major portion of the land-mobile market with an operational requirement for voice communication. Such a market segment certainly exists among public radiotelephone users, whose communication would often be directed to individuals outside of a particular company, and thus are not very predictable, nor amenable to coding. There is considerable market movement from voice to nonvoice systems (e.g., Houston Yellow Cab, the world's largest taxicab company), and none of which the author is aware from nonvoice to voice systems. The reason is almost universal agreement that nonvoice mobile systems are more cost-effective for routine business purposes. Nevertheless, it should be assumed, for scaling purposes, that half of the land-mobile market must be excluded from RDSS capture due to its lack of voice capability.

Evidence of the cautiousness of assuming that half the land-mobile market is voice-only may be found in content analysis studies of the business radio service, by far the most utilized land-mobile radio service. Such studies as that summarized in Table 4.4 show that most usage of land-mobile radios is for relaying location information and maintaining contact with dispatch centers, which, of course, are functions that RDSS can handle quite well. Functions such as conducting business, for which voice service could be important, ranked very low in terms of frequency of channel use.

One of the main reasons that private radio systems exist is that the operators of these systems do not service that which is outside of their control. With RDSS there may be many land-mobile users sharing the system, but private control can be maintained via software-defined networking. Such software-defined control can take many levels, up to and including a dedicated secondary hub with one's own set of Gold codes or other spread-spectrum encoding. With this extent of hardware- and software-defined privacy, there should not be much, if any, loss of market share due to common sharing of satellite facilities. Nevertheless, in deference to the history of private land-mobile usage, half of the market will be deemed unaddressable due to lack of *total* user control.

Table 4.4

Functions of Business Radio Communication Ranked by Category of Use
(National Survey)

Purpose	*Ranked Frequency of Use* Emergency Use	Routine Use
To provide information (including locational and operational)	1	1
To receive information (including locational and operational)	2	2
To deploy personnel	3	3
To maintain contact with office while in the field	4	4
To maintain contact between employees in the field	5	7
To deploy materials or equipment	6	5
To maintain contact with other employees in the organization	7	6
To provide technical assistance	8	8
To receive technical assistance	9	9
To conduct business transactions	10	10
To maintain contact with clients while in the field	11	11
Other	12	12

Note: Each respondent was asked to indicate the major purposes for which business radio was used by their firm. They could respond positively to more than one of the listed reasons, as well as list their own reasons. The ranking was achieved by summing the positive responses for each use. Thus, for example, more respondents indicated that they used their radios to provide information than to conduct business transactions.

Source: Bowers, R., A. Lee, and C. Hershey, *Communications for a Mobile Society* (1978), p. 154 (available from University Microfilms International, London).

It is difficult to forecast the rate at which RDSS technology might penetrate the land-mobile market. Figure 4.3 presents an array of penetration rates for various technologies of benefit to long-haul trucking, the slowest land-mobile market to penetrate. Those technologies that achieved almost immediate and total penetration were required to do so by law. It may well be the case that legislative authorities determined it advisable to mandate RDSS transmissions for specific, justifiable purposes, such as ensuring that hazardous

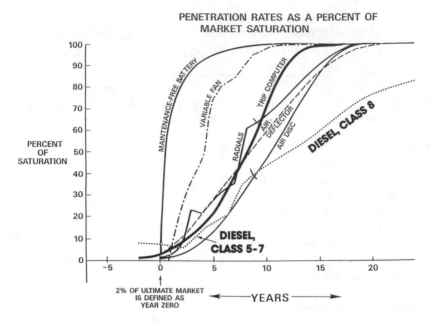

Fig. 4.3 Comparison of the Build-up Rates of Various New High-Technology Products
Source: Booz-Allen & Hamilton

chemical waste shipments follow prescribed routes to permitted dumping sites, or that speed limits are respected on transport routes.

Table 4.5 presents an expected value for the RDSS capturable share of the land-mobile communication and positioning market. An addressable market is determined according to the assumption of 50% unaddressability due to lack of voice, as described above. Potential RDSS local area market is taken from the midrange of the FCC projection in Figure 4.2. The upward rate is taken as 5% every 18–24 months, the mean of the technology adoption curves shown in Figure 4.3, and service charges are assumed to be $50 per month.

4.4 PERSONAL RDSS

Personal RDSS entails the use of a transceiver as a new personal communication or consumer electronics technology. Transceivers have several advantages over similar technology. Most of these advantages are evident from Figure 4.4.

Personal communication is a very rapidly growing industry. In 1982 the US paging population exceeded one million people for the first time. In that

Table 4.5
Expected Approximated Value for RDSS Land-Mobile Market

Parameter	2nd Year	4th Year	6th Year	8th Year
POTENTIAL MARKET (millions of transmitters)				
Long-haul	1.9	2.1	2.3	2.5
Local area	3.0	3.3	3.6	4.0
Subtotal	4.9	5.4	5.9	6.5
CAPTURABLE SHARE (%)	5.0	10.0	15.0	20.0
RDSS LAND-MOBILE PROJECTION (in millions)				
Transmitters	.24	.54	.90	1.30
Revenue (US$)	72.0	162.0	270.0	400.0

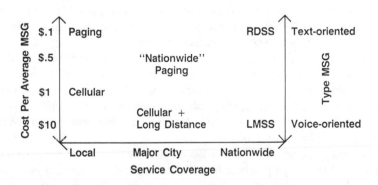

Figure 4.4 Competitive Mobile Communication Systems

year, the Yankee Group and Frost and Sullivan predicted that there would be approximately one million additional paging subscribers added each year through the end of the century. At the end of 1986, 5.5 million pagers were in use. The RDSS transceiver provides two-way service, whereas pagers only operate in the outbound direction. The RDSS transceiver provides service throughout a continental coverage area, whereas pagers only operate within the local coverage area of a land-based transmitter. RDSS transceivers seem certain to capture the high end of the pager market, and probably the low end of the cellular telephone market.

4.4.1 Personal RDSS Market Potential

The market for personal communication, essentially defined as the current paging market, is in excess of $2 billion (*circa* 1987), including both service and terminal-rental revenues. Monthly service fees run from $5 to approximately $50, depending on the sophistication of the service, with $15 being a generally acknowledged low average. Table 4.6 provides the Yankee Group's projections (1982) for paging growth to 1990, based on historical growth rates through 1987. Table 4.7 offers annual estimates of personal communication market growth, based on straightforward extrapolations from the Yankee Group's analysis.

4.4.2 Personal RDSS Competition

Competition in serving the personal communication market can be expected to come primarily from existing and new paging companies. Telecommunication analysts have not viewed portable cellular land-mobile and mobile satellite communication systems as a source of competition for personal communication because of the much higher equipment and service costs such systems require and their essentially mobile (or transportable), rather than portable, nature. RDSS includes all of the capabilities of a pager with the exception of an ability to receive signals deep within buildings such as hospitals. Approximately 10% of all pagers are part of private paging systems and used primarily within buildings. We can derive a potential personal RDSS market estimate, calibrated by the paging market, of $17 billion excluding private pagers, a figure based on 5,225,000 public paging subscribers in 1987, an annual growth rate that is predicted to decline by about 1% per year, and an average monthly service charge of $10.

To identify the addressable market on an annual basis, we must determine how many existing paging subscribers are in the market to obtain new paging service each year. The Yankee Group has estimated that 10% of pagers are lost each year and 4% are damaged beyond repair. If we assume that 30% of these are insured for replacement, then about 10% of the previous year's paging population would be in the annual addressable market due to loss of their paging number. Because paging technology is changing rapidly, it is unlikely that pagers have a useful life of greater than four years on average. Hence, the number of four-year-old pagers would also constitute part of the annual addressable personal RDSS market due to equipment obsolescence.

Table 4.6
Common Carrier and Private System Paging Market

Type	1982	1983	1987	1990
RCCs:				
Subscribers (millions)	1.38	1.69	3.7	7.25
Annual Growth (%)	—	22.5	21.7	25.1
Number of Systems (thousands)	3.775	4.7	7.5	10.85
WCCs				
Subscribers (millions)	0.35	0.42	1.06	2.38
Annual Growth (%)	—	20.0	26.3	31.0
Number of Systems (thousands)	1.51	1.8	2.25	4.34
PRIVATE SYSTEMS				
Number of Pagers (millions)	0.58	0.66	1.12	1.7
Annual Growth (%)	—	13.8	13.9	15.0
Number of Systems (thousands)	9.815	11.7	15.25	15.81
TOTAL				
Installed Pagers (millions)	2.3	2.77	5.88	11.33
Annual Growth (%)	—	20.4	20.7	24.4
Number of Systems (thousands)	15.1	18.0	25.0	31.0

Note: The estimates are from a projection by the Yankee Group of the growth in all types of paging systems. RCC stands for *radio common carrier* and is the largest subclass. WCC includes paging systems operated by regional and independent telephone companies. The private systems are those operated within private industry by firms for their own use.
Source: The Yankee Group (1982).

Table 4.7
Personal Communication Market Forecast
(Based on Paging Industry)

Year	Subscribers (millions)	Annual Service Revenue (millions US$)	Percentage of US Population
1986	4.4	526	1.8
1987	5.3	637	2.2
1988	6.4	748	2.6
1989	7.6	892	3.0
1990	9.1	1066	3.6
1991	10.7	1257	4.2
1992	12.5	1471	4.8
1993	14.4	1706	5.5
1994	16.6	1962	6.3
1995	18.9	2237	7.1
1996	21.3	2528	7.9
1997	23.8	2832	8.7

4.4.3 Personal RDSS Market Share

An estimate of the RDSS share of the personal communication market may be derived from an assumption of generally equal market division among various types of paging service and personal RDSS. The basic types of paging service are tone-only, tone-voice, and alphanumeric. Due to the low cost of tone-only, it will be assigned a weighting factor of 2. This implies RDSS may ultimately capture 20% of the personal communication market.

On the basis of competitive considerations of Section 4.4.2, a maximum market share of 20% and a standard S-curve growth, which peaks in eight years, the personal RDSS market may be estimated as forecast in Table 4.8.

It is in the area of personal RDSS that questions of privacy most often arise. No one feels comfortable thinking that someone else could track their location throughout the week. The author believes that such fears are misplaced, however.

First, commercial operators of RDSS systems will be market-driven to uphold the privacy rights of their customers. This may well mean strict adherence to a policy of not releasing the position or message traffic of any

Table 4.8
Personal RDSS Market Forecast

Parameter	2nd Year	4th Year	6th Year	8th Year
Personal communication market share (%)	5.0	10.0	15.0	20.0
New personal RDSS transceiver (thousands)	100	290	661	1,050
Cumulative RDSS transceivers (thousands)	140	570	1,450	2,940
Personal RDSS revenue (millions US$)	23	108.0	267.0	442.0

user without that user's written consent. Second, such commercial operators would probably not find many customers who wanted to pay for their positions to be relayed to someone else. An obvious class of exceptions are the industrial customers that are subscribing to the service so that headquarters will know where its employees in the field are at any time. Finally, an RDSS subscriber can always turn his transceiver off! The potential for privacy invasion is not much different with RDSS transceivers than it is, for example, with credit cards in our modern society.

4.5 SPECIAL RDSS APPLICATIONS

Special RDSS applications are those which use the technology's capabilities to accomplish a function significantly more complex than that of general position determination, position reporting, or communication. These special applications combine the positioning and messaging capabilities of RDSS to perform a physically decentralized, large-scale activity. Most of these applications apply to activities that are traditionally governmental functions.

Special applications surveyed in this section include law enforcement, demography, hazardous material control, and international liaison. For each such field of application this section provides a description of (a) the *market demand* or *requirement*, (b) the RDSS system *expression* or *scenario* that meets the demand or requirement, and (c) the *socioeconomic issues* that are triggered by the particular use of RDSS technology.

4.5.1 Law Enforcement

The law enforcement function applies position determination and communication to the complex task of ensuring compliance with a diverse set of laws and regulations over wide geographical areas. It is an axiom of modern urban life that the bigger a society becomes, the greater becomes the need for formal law enforcement procedures and it is axiomatic to police science that law enforcement must adapt to the criminal environment. If offenses are committed in moving vehicles, then the police must have cars. If drug dealers use electronic communication (pagers are a standard tool for urban drug runners), then so, too, must the police. These principles help explain the great emphasis today on law enforcement and telecommunication technology.

The scale of society grows continually toward the global limit. The ratio of rural to urban dwellers is, in technologically developed countries, the reciprocal of its value only one hundred years ago. That value was over 10! Towns merge into cities, which merge into nation-like metropolises. Countries evolve ever-greater transborder economic and political integration, which presage global organizational structures. Cultural traditions, which effected police functions when the scale of society was small, fail to work on the macroscopic level. For example, the incentives for honest dealing are lessened when people are involved with those whom they are unlikely to meet again. Law enforcement is thus needed to manage the consequences of anomie in modern large-scale society.

Communication technology has always been a horn of plenty for the law enforcement community. The first general use of land-mobile radio communication was by police, *circa* 1930, with bulky installations in the earliest patrol vehicles. In the 1950s, the development of the transistor enabled major improvements to World War II "walkie-talkie" technology, leading to their widespread use in police forces as a supplement to fixed telephones on a policeman's beat. Today, a modern police vehicle incorporates portable and mobile ("jerk and run") half-duplex voice communication equipment as well as a direct half-duplex vehicle-to-computer circuit. The purpose of this technology has been to enable the enforcement command center to improve the following:

- Receive and process police calls from the public;
- Audit and dispatch police resources in response to police calls;
- Receive and respond to information inquiries as well as calls for assistance from police patrols;
- Maintain moment-to-moment status information on police resources for command and control purposes.

The ability of technology to enhance traditionally centralized enforcement functions played a major role in the professionalization of police science. Like other professions, technology has now enabled law enforcement practitioners to contribute new knowledge about their field in systematic written form, to become subject to the norms of professional societies, and to require continuing formal education. Because RDSS accomplishes all four of the items above in an integrated fashion, it is likely to become a standard tool for the law enforcement professional of the 1990s.

4.5.1.1 Requirements for RDSS in Law Enforcement

Three significant requirements for RDSS techonlogy within law enforcement activities are (a) *command, control, and communication* (C^3, pronounced "see-cubed") for dispersed law enforcement resources, (b) increasing the percentage of *citizen calls* for police help that result in arrests, and (c) *relieving prison congestion* through electronically monitored nonincarceration of the activities of convicted individuals. While many other requirements for RDSS in law enforcement exist, these requirements exemplify the "prevention-arrest-rehabilitation" span of law enforcement activities.

The need for C^3 in police and military activities is a consequence of centralization of responsibility. With centralization comes the need to issue instructions to dispersed officers (*command*), to monitor compliance with the issued instructions (*control*), and to obtain data on the effect in the environment, whether "uptown" or "battle area", of the instructions (*communication*). For example, a standard police dispatch routine is to *command* a vehicle to a reported robbery scene, ensure through *control* that the vehicle gets there promptly, and *communicate* whether in fact a robbery is in progress. The process begins again with further *commands*.

A second major demand among law enforcement agencies is to increase the percentage of citizen reports of crimes that result in arrests. Exhaustive research in the 1970s under grants from the US Law Enforcement Assistance Administration showed that the time delay between the occurrence of a crime and reporting was the largest single factor accounting for low arrest rates. Once the location of a crime was reported, urban police generally arrived with dispatch. The lack of a general pedestrian ability to report a crime in progress — an inability that RDSS remedies — undermines the ability of police to apprehend the suspected criminal.

A third major requirement of the law enforcement system is to rehabilitate criminal elements. This requirement is well reflected in the statistic that over half of all felonies are committed by persons with a previous criminal record. There is now general recognition that prisons are not rehabilitative for most people. Individuals naturally reflect the characteristics of their social environment. The social environment of prisons apparently encourages crime. Rehabilitation requires placing individuals in a social environment that is nurturing and supportive. RDSS may fulfill a pressing demand for technology which can efficiently ensure that large numbers of nonincarcerated individuals are remaining in a prescribed location and complying with their parole conditions.

The reasons for an offender-monitoring technology are based primarily on the high cost of building and maintaining new jails and prisons. It can cost over one hundred dollars per day to house and feed an offender in prison. If that cost can be reduced by using electronic monitoring, governments can be expected to adopt such systems. Also, for every jail and prison inmate, there are three offenders serving sentences in the community. An offender-monitoring system may well pay for itself if it increases the effectiveness of compliance with parole, probation, and other community-release programs.

4.5.1.2 Law Enforcement RDSS Implementation Scenarios

Law enforcement RDSS will likely be implemented by using many unique software-defined private networks that derive transmission capability from one or more common national or multinational satellite systems. The private networks would be design-optimized for different sets of requirements, such as C^3, general emergency use, and rehabilitative applications. There is considerable flexibility with RDSS technology as to the extent of software and even hardware that the system's user controls *directly*.

With regard to C^3 applications, we may anticipate that RDSS transceivers will be issued to all law enforcement field personnel, either as stand-alone devices or integrated with a voice-band portable radio. A separate RDSS transceiver may well be built into the organization's land vehicles, sea vessels, and aircraft. The positions of these transceivers would ordinarily be automatically reported via the space segment at uniquely determined periodic intervals. The organization's central authority would obtain this position data at some point in the RDSS control-segment process from RF demodulation to value-added data processing. A baseline case would call for all data packets with transceiver ID numbers registered to a law enforcement organization to be directed into an outbound high-speed channel to that organization, where the data (vehicle or person, time, position, and message) could be stored, displayed, and distributed by any means that the software engineers can handle.

If the scenario just described is in place, then, for example, a command can be given via RDSS for several field officers to go to a certain location. The command itself could be limited to those officers near the desired location due to the commander's possession (in his computers) of recent location data on his field personnel. In the time that passes after the command is given, a location readout on his or her directed field personnel gives the commander feedback as to execution of the order. The command can be controlled by taking action to account for variances, with such variances from commander intent being displayed in real time on a computer screen. Data and codes communicated from the field provide the commander the information he or she needs for giving further direction.

With regard to general public use of RDSS in emergency situations, the objective is that of increasing the probability of capturing criminals by reducing the time between criminal act and police report. It is, of course, unrealistic to expect consumers altruistically to employ RDSS for this purpose. The question is instead one of what type of software-defined network is possible with RDSS that accomplishes the desired crime-reduction application without asking consumers to purchase RDSS for a crime-reduction-related purpose.

One concept relating to general public assistance to law enforcement efforts by use of RDSS involves a dominant role for the municipal government, which is the entity that bears most of the financial cost of combatting criminal activity. In this concept, the municipal government and an RDSS service reseller publicize a program whereby reports via RDSS of criminal activity in progress are anonymously rewarded with a one-year credit against RDSS service charges to the transceiver ID number that made a report resulting in arrest. The municipality should benefit through a significant reduction in crime and its attendant human, police, and judicial-system costs. This reduction in crime would occur, according to the studies funded under the Law Enforcement Assistant Act, due to the reduction in time between a criminal act and a report to police. The success of 911 systems provides evidence of this connection. The RDSS service reseller benefits through the support of the municipal government for using and carrying an RDSS transceiver. The reward would be set low enough to avoid "bounty hunting" or excessive false alarms, but high enough to capture public attention.

The software design for an RDSS service based on the 911 concept would probably deliver all emergency messages to the local public safety organization, which would itself be responsible for the software design of its emergency-response system. One use of the RDSS capability would involve automatic triggering of outbound RDSS emergency messages, with a location to which to go for those closest public safety personnel within a specified radius of the transceiver, being that which gave rise to the emergency transmission. For example, imagine a pedestrian assault within view of a hotel doorman equipped with a RDSS transceiver. As the pedestrian assault gave underway, the doorman transmits an emergency code via his RDSS transceiver. This transmission hits the municipality's RDSS user computer about one second later, where a computer routine determines the closest five RDSS-equipped police vehicles to the doormans geographic coordinates. An outbound "emergency response needed" message is automatically formatted and addressed to the RDSS transceiver ID numbers of the five police vehicles, and it is received by those vehicles in less than a second. In the meantime, a human operator at the municality'emergency-response center begins to question the doorman about the details, and telephone contacts are made. The contribution of RDSS is highlighted by the fact that in much less time than it takes to complete any criminal act, the police vehicle closest to the criminal act has been alerted that an emergency is underway.

With regard to rehabilitative applications, the use of RDSS to monitor the location and activities of nonincarcerated convicts may lead to an easing of prison crowding and greater success of rehabilitation. Miniature RDSS transceivers, perhaps worn like jewelry, would be used in these applications to report the location of individuals subject to the justice system. The transmission of some data over these or other transceivers would enable other individuals, such as family members, a local counselor, or an employer, to report on the convict each day. As of early 1987, more than 50 programs in 20 different states were using land-based radiodetermination systems to monitor the location of some 1000 offenders under probation, parole, pretrial release, and work-release sentences. Unfortunately, although improvements are being made, these VHF systems have proved to be relatively unreliable technically. It remains to be seen if RDSS would be more effective.

In general, offender-monitoring use of RDSS would call for the offender to transmit at some time each day, perhaps as often as every hour. His or her location is determined via the RDSS system. When placed on the RDSS system, the justice system officer could program different geographic constraints into the user software associated with each offender's transceiver ID number. If the offender's location is outside the predetermined permitted geographical area, an indicator would attract the attention of a human operator for initiating follow-up action.

4.5.1.3 Social Issues of Law Enforcement RDSS

Social issues of law enforcement RDSS primarily arise with regard to the enduring issue of public *versus* private rights. The public right is to maintain a safe society, and pursuant to this right, invasive steps may be taken against the privacy of individuals who are reasonably suspected to have engaged in criminal activity. The private right is to maintain respect for one's dignity as a human being, including the right to be free from unwarranted searches and seizures of either physical property or personal attributes. This issue is presented most graphically in the case of forcibly monitoring the location of individuals with criminal backgrounds via personally affixed transceivers.

The continual rise in prison populations has created massive overcrowding conditions. This, in turn, undermines even minimal conditions of human decency. If enforced monitoring of convicts via RDSS invades personal rights, it is arguably a far lesser invasion than that imposed by prison life itself. However, it may be said that if mandatory electronic monitoring of individuals is permitted at all, it will soon encompass more than only those persons released from prison. Consequently, a permissive view of the technology will lead to its being readily applied to individuals who would not otherwise be convicted of a crime.

An extreme use of RDSS technology could conceivably mean real-time position monitoring of all individuals in some defined community, including perhaps an entire national population. To accomplish so widespread a task, however, RDSS technology would need to be complemented with indoor-oriented electronic systems. For instance, doorways to buildings or parts of buildings could incorporate radio-scanning devices that trigger identification-readout transmissions from discretely worn transmitters (e.g., anklets, dogtags, belt packs). Such scanned information is then sent over a local network to a management information center, which processes, organizes, and reports on the data. Indoor card-access systems, whereby magnetic identification information on a card is read and checked before doors to building sectors are unlocked, have been in common use since the mid-1980s. The potential of RDSS is only to extend the range of such systems to the outdoor environment and to provide a common, nationwide, high-capacity channel for a large-scale network.

The fact that RDSS does not in itself enable a nationwide person-monitoring system, raises the question of whether RDSS is even necessary for a nationwide person-monitoring system. The answer is no. Authoritarian societies are recurring examples of the amount of monitoring of persons that can be done without any sophisticated technology. Hence, as with all technology, social choices determine the uses for which to apply new capabilities. The privacy concerns underlying such social issues are not unique to RDSS. They are part of the much larger issue of privacy rights in a high-technology society. RDSS is but one of many examples available to debaters on all sides of this issue.

To summarize, use of RDSS for law enforcement applications, such as offender-monitoring, should not serve as a "Trojan Horse" for its implementation in ever-broader sectors of society. So long as the public places a high premium on privacy in general, RDSS is no more likely to erode such a societal value than are many other technologies. When RDSS does clearly affect rights of privacy, such as the right to be secure from an unreasonable search and seizure of one's personal location, the social inquiry will focus upon *reasonableness* of the invasive activity. So long as the RDSS-based invasion of privacy is less than a supplanted invasion, such as imprisonment, no social values can be offended logically. Also, in the case that there is reasonable cause to believe that the person being monitored is a menace to the public, such as in the case of a suspected criminal activity, an RDSS-based nonconsensual monitoring of the person will likely be considered even more reasonable than a telephone wiretap, and perhaps only slightly more invasive than traditional in-person surveillance.

4.5.2 Demographic RDSS

Demography is the quantitative study of populations. RDSS has considerable promise for making quantitative population studies more comprehensive, accurate, and economical than under current techniques. Censuses are the most fundamental tool of demographers. A census taken via RDSS provides a means of automatically *geocoding* population and economic data. RDSS also enables *real-time analysis* of demographic data at minimal cost, a feature with broad application in commercial demography.

4.5.2.1 Requirements for RDSS in Demography

Requirements for geocoding population data have been fundamental to censuses since the earliest times. In fact, rough-scale geocoded maps are the starting point for the enumeration process and the benchmark for judging comprehensiveness. Legislative and private rights, such as water rights, often evolve on the basis of geocoded demographic knowledge. Public resources, such as program funding, are often most equitably distributed when based on accurate geocoded data. In every society, there are few bonds as strong as those between a family and its land.

Requirements for real-time analysis of demographic data arise as important logistical concerns for census-takers and as economically valuable information for certain commercial or governmental groups. For the 1990 US Decennial Census, over $5 billion and 200,000 enumerators will be used in a multiyear effort to reflect and accurately geocode the wide variety of demographic characteristics of 250 million Americans. Particular emphasis is placed on capturing inventory and respondent data in machine-readable form as early as possible in the enumeration process. This makes the process more cost-effective than otherwise possible, and enables earlier corrective action to cover overlooked areas or discrepancies. RDSS is an ideal technology for generating geocoded digital data at the earliest possible point in the enumeration process and for relaying such data in real-time to census managers. In general, RDSS addresses each of the following three imperatives in the field of demographic data generation:

- Greater acceptance of the idea of countrywide enumerations for scientific and governmental purposes (i.e., facilitates compliance with census enumerators);
- Improvement of administrative techniques of enumeration, including legal safeguards ensuring that individual answers will be kept confidential (i.e., enables most efficient use to be made of enumerated data);
- Deepening and systematization of the types of information obtained (i.e., optimizes the kind and form of data that enumerators acquire).

Commercial demography is now fundamental to national marketing and public opinion polling. Advertising is often made a function of geography as well as demographic characteristics, and distribution is a geocoded function. (By analogy, as the Teamsters union has long reminded, virtually everything gets to us at some point via truck.) As accurate as possible an understanding of a service or product's market is mandatory to make the most efficient use of marketing budgets. The aspect of RDSS that provides real-time knowledge of data and geography, with data being consumer response, provides a new means of obtaining dynamic understanding of large-scale markets.

Commercial demography and public opinion polling cross paths via interactive television systems. With such systems, or ones with similar capabilities that rely on telephone links, it is possible to measure dynamically viewer preferences for broadcast programming, mail-order products, and public policy matters. The prospect of electronic elections, with voting from homes, is no longer considered to be a far-fetched idea. With RDSS, commercial demography and public opinion polling will find a new means of obtaining dynamic data. In the case of the electronic elections, for example, voting by RDSS need not be from the home and registration checking can become a highly automated process.

4.5.2.2 Demographic RDSS Implementation Scenarios

Two types of demographic RDSS network scenarios are likely, one for official censuses, another for applications that depend upon general public use of the technology. For official censuses, it is likely that the government would

contract with an operator of RDSS for bulk use of transmission and positioning calculation services over a period of time. Most of the transceivers may
well be rented because they are not used continuously. At the census headquarters, computers would load transceiver ID, time, enumerator data, and
associated geographic positions from the RDSS operator's computing center.
These data would then be further processed, organized, stored, and reported
in accordance with unique census agency objectives.

It is also possible that census agencies may wish to obtain the raw time
differential information from the RDSS operator. This enables the census
agency to make an independent determination of geographic positions, perhaps
with greater accuracy and less expense than would be the case if the RDSS
operator's computers performed this work. Accuracy of geocoded data is of
continuing concern to demographers. Indeed, a significant driver of demand
for RDSS among census managers has been the pervasive lack of consistency
between where dwellings appear on supposedly accurate maps and their
precisely surveyed coordinates. Even if the differences are measured in meters,
the consequences of disputes and their volume in a large country make special
efforts to enhance geocoded accuracy quite worthwhile. The uniformity and
high accuracy of RDSS are among its most attractive features.

With widespread public use of RDSS, the transceivers will likely become
a media for effecting electronic fund transfers. One or two buttons pushed in
addition to someone else's transceiver address could effect transfers of money
from one party to the other through pre-existing arrangements between the
RDSS service company and each party's financial institution. Such arrangements would have been established according to requests from either the RDSS
subscribers *or* the financial institution's customers.

Another commercial demography scenario entails the sale of geocoded
advertising "*space*" by the RDSS operator to companies with geographically
dispersed retailing operations. The advertising "space" constitutes the periodic
(50-100 times per second) interrogation downlink signal, at 2492 MHz, with
its several-millisecond duration. Rather than the control center sending out
a default (from the user transceiver's perspective) message such as "Link OK,"
it could repeatedly transmit a brief advertising slogan such as "Fuel Up at
XYZ." It is also possible for the default display to rotate slowly through several
messages, each on display for five seconds or so (e.g., "Eat Now at ABC"; "Link
OK"; "Fuel Up at XYZ"; "Eat Now at ABC").

A dynamic extension of commercial demography in marketing would
trigger specially addressed advertising messages based on the geographic
location of an RDSS subscriber. For example, suppose that the RDSS
operating company established a class of service that included lower service

charges in exchange for permitting mass marketing geocoded software to access only the geographic coordinates from each of the subscriber's RDSS transmissions. Many subscribers could find this class of service beneficial because the privacy invasion here is strictly theoretical, not dissimilar to the junk mail we get as part of a listed street address. The lower service charges would be compensated by bulk payments to the RDSS operator from a marketing company for the right to access real-time geographical coordinates with a special program that sends out advertising messages to different transceivers based on their coordinates. An exemplary message might be "XYZ Fuel 2 km Ahead."

4.5.2.3 Social Issues of Demographic RDSS

Social issues inevitably come to mind with demographic RDSS. This is due to both the potentially privacy-invasive aspects of RDSS and the apparent power being vested in the hands of the RDSS operator. Policy-makers are to decide how much, if any, additional invasion of privacy is likely with RDSS as compared to other mass electronic systems in modern society. A similar question arises with respect to the apparent power of the RDSS operator as compared to that of telephone companies, bank card companies, credit companies, and various government agencies. Generally, it may be that all operators of mass electronic systems face a common threat of system abuse, in which case, some basic regulation tends to satisfy the public's concerns.

An important feature of RDSS to keep in mind when discussing these social issues is that multiple RDSS operators in the same geographic area are fundamental to the technology's system architecture. Competitive alternatives often comprise one of the surest structural means to control offensive practices by any particular medium. Any RDSS operator that acquires a reputation for providing poor service, or engaging in activities that offend subscribers' sense of privacy, will quickly lose customers and come under financial pressure to change its practices.

The existence of marketplace competition does not really address political concerns that sensitive national data not be transmitted outside of a country's borders. These concerns are often considered under the rubric of "national privacy," and have led to much regulation of transborder data flow. It is clear that the RDSS operator could collect a vast reservoir of sensitive data. The RDSS-generated map of a country's transportation activity alone is likely to be considered salient to "national security" interests. Hence, at least on the question of international transmission, some regulation of nationally geocoded demographic data may be inevitable.

4.5.3 Hazardous Materials Control

The field of hazardous materials control encompasses regulatory and operational regimes specifically designed to control the menace to public safety inherent in the transport of certain substances. Fissionable materials and toxic chemicals are major targets of these regimes. Knowledge of the location of these shipments at all times and of any variance with approved routes and estimated time of arrival (ETA) has great value to regulatory authorities. The ability of RDSS to meet much of this requirement is likely to make it an important tool in controlling transportation of hazardous materials.

Generally, different regulatory authorities are responsible for different categories of hazardous materials. In the United States, responsibility is split largely between the Department of Energy (fissionable materials) and the Department of Transportation or the Environmental Protection Agency (toxic chemicals, including toxic industrial waste). Local governmental authorities may often exercise as much influence as federal bodies. The greater perceived threat associated with fissionable materials has led to more operational-control oriented regimes in most countries, as compared with passive-control in the case of toxic industrial waste. The Bhopal-Union Carbide industrial disaster, with its tens of thousands of victims, graphically illustrates the actual threat associated with nonfissionable materials.

In an operational-control oriented regime, regulatory authorities may be expected to take actual administrative control of the transportation process, either with employees or contractors. As a rough approximation, one full-time or reserve vehicle is needed on the road per fission energy plant year. This implies several hundred over-the-road transport trucks, not including support vehicles, in each of the United States, Europe, and the Soviet Union. Regimes responsible for the transportation of this output display a motivation to possess the most advanced and comprehensive radiodetermination techology. High-frequency radio direction-finding may be employed, but its advantage of broad coverage is offset by its bandwidth and propagation reliability limitations. System redundancy is likely to be considered cost-effective for an operational-control regime.

The position determination and reporting capabilities of RDSS, coupled with its two-way data communication nature, may well justify the use of direct RF downlinks of spread-spectrum signals by an operational-control regime. The expense and technical complexity of the RF hub may well be offset by the value attached to more secure and reliable control of a dispersed fleet. Conversely, demodulated and encrypted timing and other data could be transferred to the vehicular fleet controller. This slightly greater involvement of the RDSS operator does not diminish the user's control over position-determination accuracy and message integrity. It may also be cost-effective simply to contract for customized vehicle-tracking service from an RDSS operator, as either an auditing or an active system.

Passive-control regimes tend to be responsible for the transshipment and storage of toxic but nonfissionable industrial materials, including both intermediate and waste products. In these regimes, a body of regulation specifies what action industrial firms may or may not take with specified classes of hazardous materials. Violation of these rules is rarely taken as seriously as would be the case for interference with the activities of an operational-control regime. Formal requests for compliance or fines are the normal coercive measures. In highly industrialized countries, noncompliance with environmental regulations resulted in significantly more serious penalties in the 1980s. Industry-wide programs to rectify the improper dumping of hazardous materials, such as the US EPA's "superfund," also reflect the increasingly sophisticated approaches being employed by passive-control regimes.

An interesting implementation scenario for a passive-control regime would entail mandatory carriage and use of RDSS transceivers, wherein the positioning data are transmitted to the regime or its contractor for monitoring compliance with prescribed hazardous material routes and delivery points. Regulations often restrict the trucking of hazardous materials to roadways located away from residential areas wherever possible to minimize the risks of an accident causing widespread damage to health and safety. Compliance with such restrictions may not hold incentive for carriers, perhaps because such routes are longer. At present, there is very little in the way of route compliance monitoring. Indeed, non-RDSS monitoring could prove a mammoth task. In the United States, up to 300,000 vehicles become involved in transporting bulk cargo subject to EPA route regulation. Mandatory RDSS reporting would accomplish the task of ensuring that the law is followed and the potential of public health disasters are kept to a minimum.

Social issues relating to hazardous materials control and RDSS are of an indirect nature. The main question is whether RDSS encourages a "false sense of security" about an activity, hazardous materials transportation, that should be minimized as a matter of public policy. We may recall that RDSS must be coupled with other systems to work in areas where line of sight is absent, such as tunnels, and is subject to nefarious designs as are many of the most advanced modern technologies. Nevertheless, to the extent hazardous materials transportation is a fact of industrial society, the availability of RDSS is certain to be welcomed as at least a redundant or monitoring system.

4.5.4 International Liaison

International liaison refers to both formal and informal networks of coordination between entities in different countries. The concept is very similar to international communication, except that international liaison excludes mass communication, such as broadcasting, and it is specifically oriented toward intercultural communication. The readily available transborder capability of RDSS, coupled with its *digital syntax*, may well lend itself to general international correspondence. International commerce may also be facilitated by using RDSS as an "electronic bill of lading" medium. Finally, to illustrate the scope of possible application of the RDSS technology, it may be possible to employ radiodetermination monitoring international arms control agreements.

Translingual Dialogue

RDSs may be said to employ a digital syntax, which facilitates international liaison. This results from the burst-mode form of modulation of RDSS transmission, with data packets of about 32 to 128 bytes in length. Messages or information must be "packed" into this format, which establishes much *uniformity* (hence, "digital") and may be heuristically considered as a kind of *grammar* (hence, "syntax"). This syntax lends itself quite readily to automatic computer translation between languages.

With the appropriate *language translation selector* reserved from the RDSS control center, it may be possible to carry on a quasiconversation or electronic correspondence between two or more persons speaking different languages. The computer would have matrix tables that mapped words in different languages to their corresponding words in other languages. The portability of RDSS enables someone with a transceiver in hand to access as much knowledge as is made available through the control center, perhaps including translation capability among dozens of different languages. Although ambiguity would always exist, and hence translation would be problematical, there may come to be an interesting new international technocultural capability that does not currently exist.

International Commerce

International economic interdependence is taken for granted as we reach the end of the 20th century, but most such commerce is still conducted through media crafted in the days of imperialism. Bills of lading, for example, are pieces of paper evidencing the agreement of a carrier to accept goods from one party and to deliver the same goods to a second party. Rights of ownership and risk of loss attach at various points in the bill of lading process. The ability of RDSS to provide continuous position monitoring and to incorporate various sensor data may combine to make it a medium of international commerce by the turn of the century. RDSS can also be used as a channel for electronic funds transfers associated with international shipments.

Industry standards are already well developed in the area of local area container identification systems, with defined data format and electronic scanning specifications in place. Identification tags are scanned at points of embarcation and disembarcation, and they are attached so as not to be damaged by the stacking of containers in a ship's hold. RDSS transceiver tags affixed to cargo containers would operate in a similar fashion, but they could provide many more services. At a point of loading, container-based RDSS transceiver tags could be discretely addressed by the control center, and their ID codes may be linked with those of the ship's RDSS transceiver. Thereafter, the container's position would be reported as that of the ship's position. At the unloading site, a second interrogation process would reactivate the container's RDSS tag as the primary positioning source.

It may well come to be that governments will mandate the use of active mobile inventory control systems such as RDSS for international commerce in goods that are subject to export controls such as militarily capable technology. For such technology the costs of losing control over its destination can be very high, perhaps billions of dollars for advanced electronics that fall into unfriendly hands. Although no control system can be made foolproof, RDSS represents a quantum leap in real-time radiodetermination and monitoring capability. Given the values involved, RDSS may well justify mandated use on shipments subject to special export regulations, such as the US International Traffic in Arms Regulations (ITAR).

International Peace Monitoring

The area of arms control monitoring is rich with exotic technology. Notwithstanding this reservoir of capability, there is some reason to conclude that a significant arms control monitoring capability can be played by RDSS. First, the requirements for arms control monitoring systems vary greatly, implying a variety of techiques. Unique systems are required for strategic weapons, tactical weapons, and certain Third World tension zones. The inherent positioning capability of RDSS combined with sensor-data reporting meshes well with generic arms control monitoring requirements. Whether the active-interrogation nature of RDSS is an obstacle depends on the particular weaponry being monitored, and it remains unclear whether direct interface between radiodetermination monitoring systems and remote sensing or imaging systems is advisable.

Implementation of RDSS as an arms control monitoring system raises the immediate question of control center responsibility. The United Nations has long been the subject of proposals for an impartial arms control monitoring entity. The United Nation's Committee on the Peaceful Uses of Outer Space has developed concepts for a directly operational UN role in outer space, such as satellite communication for peace. The UN already serves as the central international registry for all objects launched into space, pursuant to an international treaty signed in 1974. The International Liability Convention of 1971 specifies an arbitration role for the UN for multinational accidents that occur in space. Also, since the very dawning of the space age in the late 1950s, the UN has been at the forefront of those calling for accelerated development of space and earth orbits in the best interests of mankind. Hence, operation of a worldwide RDSS system for arms control monitoring, perhaps with relay payloads cost-effectively appended to several nations' host satellites, may well become the appropriate embarcation point for the UN as it accelerates its efforts to encourage the peaceful development of space.

4.5.4.1 Liaison and Society

In each of the three areas outlined above, the unique characteristics of RDSS are useful in establishing international liaison. This seems fairly obvious in the case of direct international translation, while it may be less obvious, but no less important, in the field of international commerce. Such businesses activities depend on certain standards of commercial conduct being respected on both sides of an international border. Use of RDSS appears to have much promise as part of a new set of 21st century international commercial norms. Finally, in the case of arms control monitoring, the role of RDSS is that of an "ombudstechnology": a system that is trusted by all sides to perform an objective task. RDSS, with its ubiquity and digital syntax, helps information flow between different value systems. This flow of information is the essence of international liaison, whether such value systems are represented by people, trading companies, or national defense establishments.

Societies are held together through communication. When society becomes larger in scope, communication systems must be used to weave the necessary bonds. Radiodetermination is a new resource available to society. For developing countries, RDSS can become a medium of national liaison that enables "leap-frogging" into personal communication parity with developed countries. For developed countries, RDSS enables a significantly closer wireless bond to be established among the citizenry and the economic infrastructure nationwide. The establishment and fostering of such bonds are basic objectives of all societies. Indeed, whether being applied to safety of life or efficiency of industry, RDSS forges societal communication networks that are likely to be of great value for many years to come.

Bibliography

Aeronautical RDSS

1. International Organization of Aircraft Owners and Pilots Associations, "Surveillance Capability of the Radio Determination Satellite Service," Third Meeting of the Special Committee on Future Air Navigation Systems of the International Civil Aviation Organization, Nov. 1986.
2. O'Neill, G.K., "A Multi-Purpose Satellite System to Serve Civil Aviation Needs," *ICAO Bulletin*, March 1985.
3. Radio Technical Commission for Aeronautics, "User Requirements for Future Communications, Navigation, and Surveillance Systems, Including Space Technology Applications," Report of Special Committee 155, 1986.
4. Watts, N., "Air Traffic Control, Cybernetics and Radiodetermination Satellites," unpublished monograph (available from the author), Federal Aviation Administration, Research Center, Atlantic City, New Jersey.

Maritime RDSS

5. Bell, J.C., "Standard-C and Positioning," *Proc. Conf. Royal Institute of Navigation*, 1986.
6. Bogdanov, V., "Basic Principles for Planning and Development of an International Satellite Navigation: Worldwide Navigation into the 21st Century," Royal Institute of Navigation, 1986.
7. Federal Communications Commission, *Federal Radio Navigation Plan*, 3rd Ed., Washington, DC, 1984.
8. International Hydrographic Bureau, *IHO Standards for Hydrographic Surveys*, 1986.
9. Kies, P.J., "Evolution of Position Locating Systems on Data Buoys," International Symposium on Marine Positioning, Reston, VA, Oct. 14–17, 1986.

10. Maritime Safety Committee, Reports of the Radio Communications and Safety of Navigation Subcommittees, International Maritime Organization, 1985–1986.
11. Pasquay, J., "Hydrographic Requirements for Modern Shipping," *J. Navigation*, Sept. 1986.
12. Radio Technical Commission for Maritime, Working Group Reports, Special Committee 108, March 1987.
13. Sampson, S.R., "A Survey of Commercially Available Positioning Systems," *Navigation*, Summer 1985.
14. Federal Communications Commission, *Report and Order*, Docket No. 84-689 (Frequencies for Radiodetermination Satellite Service) and Docket No. 84-690 (Rules Regarding Radiodetermination Satellite Service), Washington, DC, 1986.
15. Geostar Corporation, Communications Responder Specification, Request for Blanket Licensing Authority, Federal Communications Commission, Docket No. 84-690, Washington, DC, 1986.
16. Geostar Corporation, System Compendium, Request for Operating Authority, Federal Communications Commission, File No. 2191-DSS-P/LA-83, Washington, DC, 1985.
17. International Telecommunication Union, Advance Publication, USRDSS East, Central and West, Geneva, 1984.
18. McCaw Cellular Communication, Request for Operating Authority, Federal Communications Commission, File No. 1236-DSS-P/L-85, Washington, DC, 1985.
19. Mobile Communications Corporation of America, Request for Operating Authority, Federal Communications Commission, File No. 1232-DSS-P/L-85, Washington, DC, 1985.

RDSS Positioning

20. Kalafus, R., J. Vilcans, and N. Knable, "Differential Operation of Navstar GPS," *Navigation*, Fall 1983.
21. Mertikas, S., D. Wells, and P. Leenhouts, "Treatment of Navigation Accuracies: Proposals for the Future," *Navigation*, Spring 1985.
22. O'Neill, G.K., "Satellite-Based Vehicle Position Determining System," United States Patent No. 4,359,733, 1982.
23. Richards, R., "Radiodetermination Satellite Accuracy Analysis," IEEE Position Location and Navigation Conference, 1986.
24. Snively, L., and W. Osborne, "Analysis of the Geostar Position Determination System," 11th AIAA Communications Satellite Systems Conference, 1986.

RDSS Communications

25. Bowers, R., A. Lee, and C. Hershey, *Communications for a Mobile Society: An Assessment of New Technology*, Sage Publications, Washington, DC, 1978.
26. Cook, C., ed., *Spread Spectrum Communications*, IEEE Press, New York, 1983.
27. Dixon, R., *Spread Spectrum Systems*, John Wiley and Sons, New York, 1984.
28. International Telecommunication Union, *Recommendations and Reports of the CCIR*, Vol. VIII, Mobile Services, 1986, especially *Report 1050* (Technical and Operational Considerations for a Radiodetermination Satellite Service in Bands 9 and 10), *Report 509-3* (Signal Quality and Modulation Techniques for Radio Communication and Radiodetermination Satellite Services for Aircraft and Ships), and *Report 770-1* (Technical and Operating Considerations for a Land-Mobile Satellite Service Operating in Band 9).
29. "Mobile Radio Communications," *IEEE Comm.*, Feb. 1986.
30. Roddy, D., *Electronic Communications*, 3rd Ed., Reston Publishing, Reston, VA, 1984.

Regulatory Considerations

31. Codding, G.A., and A.M. Rutkowski, *The International Telecommunication Union in a Changing World*, 2nd Ed., Artech House, Norwood, MA, 1987.
32. Federal Communications Commission, Private Land Mobile Requirements, *Final Report*, Washington, DC, 1985.
33. Federal Communications Commission, *Second Report and Order*, Docket No. 84-1234 (Mobile Satellite Proceeding), Washington, DC, 1987.
34. Rothblatt, M., and S. VanTill, "Radiodetermination Satellite Service and the Process of International Recognition," *Space Communication and Broadcasting*, March 1986.

Index